PATHS TO POWER

A WOMAN'S GUIDE
FROM FIRST JOB TO TOP EXECUTIVE

NATASHA JOSEFOWITZ

ADDISON-WESLEY PUBLISHING COMPANY

Reading, Massachusetts • Menlo Park, California
London • Amsterdam • Don Mills, Ontario • Sydney

The poems are very personal expressions of my thoughts and feelings, and I am happy to have the opportunity to share them. I hope they will not be judged by their poetic merit but by the echoes and smiles they may produce.

Library of Congress Cataloging in Publication Data

Josefowitz, Natasha.
    Paths to power.

    Includes index.
    1. Vocational guidance for women.   I.   Title.
HD6058.J67   650.1′4′024042   79-28721
ISBN 0-201-03485-9
ISBN 0-201-03486-7 pbk.

Second printing, October 1980

ISBN 0-201-03485-9-H
ISBN 0-201-03486-7-P

ABCDEFGHIJ-DO-89876543210

To My Family
Ancestors, Contemporaries, Descendants
yesterday, today, tomorrow

Preface

This book is about how women can succeed in today's organizations. It is *not* about how organizations should be or how women should change them. It *is* about how we women can empower ourselves and others. It is addressed to women who wish to become more knowledgeable, more effective, and better prepared to face the roadblocks to power that yet exist.

It is my hope that as more women accede to positions of power, they will remember those in low-paying, dead-end jobs and be committed to changing the lot of all who still experience discrimination. The women who have made it to the top have the privilege and the responsibility to create opportunities for others.

Introduction

Whether you are married or single, have just graduated from school, have young children at home, or have finished raising your family, I am inviting you to come on a journey with me. This journey will touch on many aspects of a woman's life at work, from your first day on the job to retirement and everything that happens in between.

Success for you may be in terms of monetary reward, position, self-fulfillment, an integrated life, or inner peace. I do not know, but what I postulate is that in order to be the most effective person you can be and achieve what you want (whatever that is), you need to get in touch with your own power and you need to understand the power of others.

This journey is two-fold—an expedition through the vagaries of organizations to the uncharted territory of corporate power, and an inner journey to *your* core identity and personal power. *You* are the guide for the inner journey; I shall lead the other expedition. You bring "knowing," I'll bring "knowledge," and between us we'll have awareness and understanding. You will learn to cope with the world as a worker, a supervisor, a manager, a top executive; as a single woman, a married woman; as a mother, young or old.

It is important here to distinguish between the "knowing" that you bring to this book and the "knowledge" that I offer. "Knowing" suggests an intuitive astuteness, while "knowledge" means "understanding gained through experience or study." The first is derived from personal feelings; the second, from public fact. I will tell you all my "knowledge" of organizational life, which I believe will be useful to you, but I will also share with you my "knowing."

I often feel a special connection to women generally. Perhaps I feel a bond of tribal knowing. It is not clear to me what it is that all we women know, but I do believe we share an unspoken secret. That bond, that shared knowing, is a source of our support and our power.

My wish with this book is to empower you, the reader, with the knowing that is in yourself and with the knowledge of the world of work. These can help you develop the skills necessary not only to survive in this world but to master it, so that you may find peace and fulfillment within yourself, comfort in your environment, and time to care for others.

The folk wisdom of "Nothing ventured, nothing gained" is correct, if what you want to gain is different from what and where you are. I am not promoting change, I am promoting *options*. You must know what your options are before you can make an informed choice. After reading this book, you may decide that upward mobility is not for you. That's alright, because your decision is based on a more accurate assessment of your strengths and limitations and of the expectations and requirements of the business world.

The knowledge of organizational life that I share with you was garnered from several different work and educational experiences. I obtained my first real job—research assistant at the Child Development Center—while my two children were still in school, so I know the difficulties of multicommitment. At age 35 I returned to school to earn my Master's degree in social work.

After my graduation, we moved to Lausanne, Switzerland, where I taught casework and supervision and also worked as a family therapist. In addition, I conducted classes, seminars, and workshops in management training and development for administrators in such fields as health services, industry, government, and education.

Later, when my children were grown, I wanted to earn a Ph.D. so that I could obtain a permanent faculty position. I achieved that goal at the age of 50 and today am an associate professor at the College of Business Administration at San Diego State University, teaching courses in organizational behavior, conflict management, interpersonal and group dynamics, and women in management, a course I developed. I still work as a consultant to industry and service organizations and am certified as a clinical social worker. I am also accredited by the International Association of Applied Social Scientists (IAASS). In addition, I am the director of the Graduate Students Professional Development Program for the National Training Laboratories (NTL).

My background and training have taught me several important things:

- It is possible to assess organizational needs and learn the skills to fulfill those needs.

- Skills are *transferable* between fields. I used social work concepts to teach organizational behavior.

- Women know more than they think they do. We are socialized to discount our capabilities.

- Risk taking is essential for success. First, however, we need self-confidence and a positive outlook. A good motto for this is *Identify your fear, then go there!*

- Empowering others—sharing your power—increases your power. But you must have it to share it.

I hope this book will help you get in touch with your knowing and your knowledge so that you gain enough self-assurance to become a risk taker, learn the skills necessary to survive and progress in the world of work, and achieve enough power to empower others.

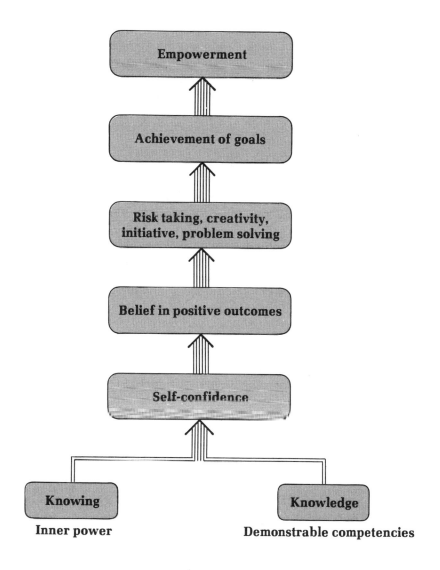

Traditionally, management books have been written by men, for men, using research data formulated about men in a male-dominated society. This book, written by a woman for women, includes new findings on the experiences of women in order to prepare other women to enter the still mostly male-dominated society. Recent research findings are just beginning to report that the nature and psychology of women are compatible with power—indeed, that certain "feminine" qualities are intrinsic to power.

This book begins with a discussion of power, then goes through the various phases of work life: looking for a job, the first days at work, the early years, supervision, middle management, top executive, and retirement. Each stage has issues particular to women, such as assertion and multicommitments, and issues pertinent to the organization, such as sponsoring, organizational development, and conflict management. It is not enough for women to identify the issues—we must identify *with* the issues in order to truly learn.

In this book you can come on board or get off at any time. Each section can stand on its own and is complete in itself. It doesn't matter if you are a young woman wondering what kind of a job to look for; a mother wondering how to juggle a career and children; a woman in your 40s returning to work after time off to raise a family; or an older woman starting work for the first time and wondering if it's too late. Come along on the journey with me!

ACKNOWLEDGMENTS

Just as this book is about stepping stones along a woman's career path, my writing it was a series of steps, influenced by friends and colleagues along the way. Rosabeth Moss Kanter, Barry A. Stein, Francine Hall, and Mildred Pryor were the first to read the early draft, and they encouraged me with their enthusiastic response. Then came Keith Nave of Addison-Wesley who believed in the book and signed the contract for its publication. Jean Baker Miller, Eileen Morley, Betty Roberts, Alice Sargent, Phyllis Schlesinger, Rita Weathersby, and Mary Winslow all took the time to write commentaries on my drafts, and I am grateful to them.

I am also grateful to my colleagues, Janice Eddy, Judith Noel, Barbara Babkirk, Phyllis Forbes, and Barbara Brockelman, who read the manuscript and whose support was much appreciated. Jenifer McKinnon, Elaine Vachon, and Geraldine Lampros typed and retyped and reassured me that I was on the right track. My thanks also go to Dean Charles Warden and to my colleagues at the Whittemore School of Business and Economics of the University of New Hampshire for providing a climate of intellectual stimulation.

Ann Dilworth, my hard-working editor, and Tess Palmer and Anne Eldridge at Addison-Wesley taught me to cut and rewrite and still enjoy the process.

Sam Josefowitz provided me with a different perspective and much encouragement. My son Paul and I had endless phone conversations about the concepts, and I thank him for his incisive comments. My daughter Nina's theory on nonassertion, from her Ph.D. dissertation, is a critical part of this book. My mother, Tamara Chapro, remains a living example of a woman's power to empower others. I am grateful to my late father, Myron Chapro, for the belief that I could become whatever I wanted.

I also want to thank all my students, who struggled with the various drafts of my book as their text and who helped me test many of the concepts.

Finally, thanks to my friend and colleague, Herman Gadon, for his unfailing support during the three years I spent writing the manuscript.

N. J.

Contents

1. Paths to Power   3

2. The Preparation   23

3. Moving In   51

4. Settling Down   71

5. A Brain to Pick, a Shoulder to Cry On, and a Kick in the Pants   93

6. Women as Supervisors   105

7. The Balancing Act: The Multicommited Woman   125

8. The Middle Manager: Managing Yourself   141

9. The Middle Manager: You and Others   159

10. Women in Groups   177

11. Women as Leaders   195

12. Top Executive   211

13. Epilog: Re-Tire to Re-Start   229

Appendixes   235

General Bibliography   283

Index   285

PATHS
TO
POWER

Chapter 1

STEREOTYPES

She said to him,
The academic life must be pleasant—
you're a professor, how nice.
He said to her
Well, maybe some day
you'll marry one.
She said to him,
Why should I marry one
when I can be one!

Paths to Power

To feel powerful! To be powerful! To act powerful!

- On a *personal* level, we *need to feel* powerful to empower ourselves and others.
- On a *professional* level, we *aspire to be* powerful to empower ourselves and others.
- On a *political* level, we *have the right to act* powerful to empower ourselves and others.

We need power in order to fulfill our *needs*, acknowledge our *aspirations*, and claim our *rights*. Power is achieved through self-awareness, through the understanding of others, and through the knowledge of organizations.

What *is* power? Who *has* it? How does it *feel*? *How* do you get it? How do you *use* it? How do you survive when you don't have it? Or don't want it?

WHAT IS POWER?

POWER! Close your eyes, say that word to yourself, and listen to what you feel. Stay with the feeling until you can describe it. Don't read any further until you have done this.

Do you derive negative connotations from the word? What are they? Does it make you sad or angry because you feel powerless? Or does the word have a nice ring to it? Does it make you sit up with pride or strength?

There are two basic ideas of power. We can take it to mean:

forcefulness:	or	effectiveness:
the ability or		the ability or
official capacity		capacity
to exercise control;		to act or perform
authority		effectively

Power as forcefulness is the traditional, finite idea of power: There is just so much, and no more, to go around. The lines of influence are vertical, going from top to bottom. The strategy is to get to the top of the hierarchy, get the largest piece of the pie, in order to have the most influence, the most impact, and the least dependence on others.

Power as effectiveness is more elastic. It can stretch, spread out, or change its shape according to the needs of the people and the organization. The word itself comes from a French word "pouvoir," which means "to be able." The word "empower" then means "to enable." This broader concept of power includes the capacity (the role) and the ability (the competence) to get things done by either influencing others or having access to resources. It also includes the idea of granting more autonomy to those with less power.

The whole issue of power resides within these two ideas. The negative connotations of forcefulness relate to furthering your own ends at the expense of others or taking pleasure in their domination, either through sexual exploitation, verbal or physical agression, or the threat thereof. This coercion and manipulation is considered to be "bad" power. On the other hand, "good" power is seen as helping people formulate their own goals and providing them with the means to achieve them. It is a concern for others, although a potential danger lies in the statement, "I do it for your own good," which may reflect personal preferences rather than a response to the other's best interests. Generally, however, "good" power makes both you and others feel more powerful.

Those with power have always tried to hold on to it; those without it, to take it away. The more power you have, the more pleasure you feel at being powerful and the more you stand to lose. The greater the differential between the haves and the have-nots, the larger the threat of takeover. But today many women (and some men) are saying something new. They do not wish to be victimized by the powerful, but nor do they wish to usurp their power. Instead, they want an *equal* place, they want to *share* the power. The only way to retain power is to share it, thus empowering others.

ROADBLOCKS TO POWER

In order to empower others, to share our influence and effectiveness, we must first have some power. The means to achieve it are the *Paths to Power*, and that's what this book is about.

Women's *Paths to Power* are very different from men's. The roadblocks encountered on the way are not the same, nor are their climbing styles similar. Men's paths take them from least powerful (most dependent) to most powerful (least dependent) on a fairly direct and well-marked road. Women's paths are not clearly marked by precedent.

The roadblocks to achieving power are both internal and external. Internal roadblocks are those of past socialization (learning what good little girls should or should not be and do), of current expectations (trying to be a Superwoman), and of daily responsibilities (handling a family and a career). The external roadblocks are encounters with prejudice, manifested in stereotyping and ultimately in societal and organizational discrimination, such as unfair hiring and promotional practices. These external roadblocks are due to socialization as well, but not ours—theirs. Most men have been socialized to stereotype women and to discriminate against them.

Women need to go through a self-empowerment process to become free of society's "you should" as well as from our own "I should."

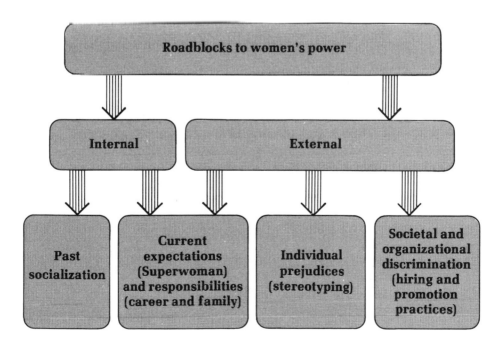

The stylistic difference in the climb to power is also important. Women must still do many of the things that most men do. But we should add awareness to all our endeavors; add caring to all our relationships; add our vulnerabilities and our capacity to share ourselves and our power; add our intuition, warmth, and tenderness; add our listening skills. If we do this, the *Paths to Power* may still have some rough spots, but there will be patches of soft moss and perhaps even a few flowers.

The Stylistic Difference

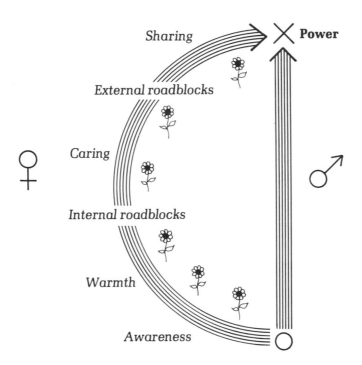

Attributed power

In order to gain power, we must also understand how it works and who has it. People attribute power to those individuals who influence behavior. A supervisor has power, for example, because the people who are supervised modify their behavior to please that supervisor. Why? Because that person controls certain resources (money, promotions, recognition, approval, information) that are valuable to them. This is realistic power. However, if you don't care about your job and don't care whether you are promoted or fired, your boss has no power over you.

Sometimes you may also attribute to certain people a power that they may or may not actually own. For instance, I attribute power to people who seem to me to have their act together. By this I mean people who appear self-confident, well-dressed, intelligent, and just the slightest bit distant or formal. To whom do you attribute unrealistic power?

Yet no one has power all the time. Power is situational. The president of a large company may be powerless with a prosecuting attorney or with the doctor who diagnoses a serious illness. Which situations make you feel powerful? Which make you feel powerless? The women in my management class gave some typical responses:

I feel *powerless* when	I feel *powerful* when
I'm ignored.	I'm energetic.
I'm on unfamiliar ground.	I'm healthy.
I'm indecisive.	I get positive feedback.
I'm exhausted.	I know I look good.
I'm sitting with an interviewer.	I tell people I have my own secretary.
I'm told what to do, without a choice.	I have clear goals.
There are too many demands on me.	I'm familiar with the subject.
I'm being controlled or manipulated.	I stick to decisions.
I have pent-up anger.	I speak out against injustice.
I feel isolated in a group.	I allow myself to be selfish without guilt.
I don't think (react) quickly.	I tell a good joke.
I have no accountability.	I'm in a supportive group.
I don't speak loud enough.	I know my expertise is greater than that of my working companions.
I feel the height difference.	I'm sitting behind a desk.
I don't have control over my time.	I ski down Tuckerman's Ravine.

THE POWER HIERARCHY

There are three basic levels to the hierarchy of power: dependent power, intermediate power, and influence power. They exist on a continuum: There is no total dependence and no complete influence.

At the bottom are those with *dependent power*. These people have little control over their lives and few alternatives or opportunities for change or growth. They are constantly influenced by others, but have little influence themselves. They feel powerless most of the time. Women, minorities, recent immigrants, children, the sick, the handicapped, the old—these are the people at the bottom of the rung. In business they usually fill the entry-level jobs.

It is important to understand that these people do not *choose* to be this way, but are so because of life's circumstances. If any viable alternatives were available, they would probably take advantage of them, provided these people had not already given up hope.

The words "dependent" and "power" may seem to be mutually exclusive, but in fact, the low-power people *do* have power. It may not be *direct* power to influence, but it is available for *indirect* use. All children are in this powerless position, but they learn very early how to influence the people on whom they depend.

As soon as babies figure out that crying will bring Mother, they have learned how to use their indirect power. By trial and error, those children soon develop a vast repertoire of strategies. They can be flirtatious or clever, they can whine or pout. They can get their way by being funny, by entertaining the adults, or by withdrawing into a sulking silence.

The same ploys are used by powerless adults. By updating their lessons from childhood, these people get what they want by cajoling, seducing, withdrawing, or sabotaging. There is little difference between telling your mother you have a stomachache so you don't have to go to school and calling in sick so you don't have to go to work.

Along the middle levels of the hierarchy are those with *intermediate power*. These people have more direct control over their own lives. Professionals such as lawyers and doctors in independent practice are usually outside the hierarchical structure and can be said to have independent power. However, they too are rarely entirely free of control from others. The laboratory researcher who works on her own still depends on grants or work that is given to her, and she is usually accountable for the results of her research. In this case the controlling forces are quite subtle, but they still exist.

A husband and wife who share their decision making and responsibilities have an equal but shifting power relationship: Although the power in such a relationship is fairly evenly distributed, one may be needier at certain times than the other. When the dependence is mutual it becomes interdependence. Interdependent power may move horizontally, as between peers, or vertically, as between managerial positions.

At the top-most level of the continuum are those with *influence power*. This kind of power is public; it is acknowledged by the people who have it and by others. The power resides as well in the *position* held, the *role* played, the *title* commanded—it is the president of the company, the executive vice president, the direc-

tor, the superintendent of schools, the dean of the college, the hospital administrator. It is the person who sits at the top of the pyramid and makes decisions with the
broadest impact on the institution and its people. Yet titles or roles are not always the
sole indicators of power. Influence may be also a matter of where you sit in the organization or to whom you are connected.

   The person who has influence power may permit others to influence her. In fact,
a good leader will allow herself to be influenced by the needs of the people. Even a
president needs people who are willing to be led. But although people with the most
influence power are, to an extent, still dependent upon others to keep them there,
that dependence has an optional quality. At the top, *you* decide to whom you will
listen and to what you will pay attention.

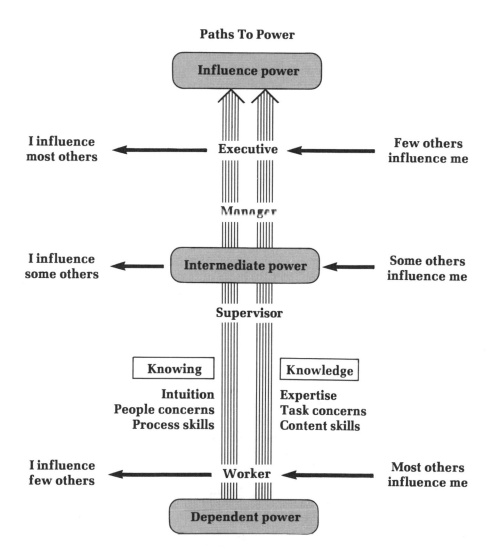

Paths To Power

Very few women have influence power because very few are in the direct power positions. However, most women *have* had the opportunity to wield influence power in their own traditional domain: Men are still the kings at work, but women rule at home, making decisions that touch the lives of the nuclear family unit. Now, however, more and more women are entering the traditional male domain in order to share in the decision making and in the influence and power. These women feel that it is equally important to be able to influence the lives of many workers. Of course, as more women share the traditional male domain, more men will have to share the responsibilities of the traditional female domain.

"Knowing" and "knowledge"

To move upward along the continuum to more influential positions, two things are necessary. First, you must acquire *knowledge*, a content orientation. Second, you must recognize your *knowing,* a process orientation. Content refers to *what* is being said. Process refers to *how* it is said. To hear the content, listen to the actual message; to understand the process, listen to the tone of voice, the nonverbal clues, the reactions of others. To have specific *knowledge* is to have expertise in some area. This can be learned. The expert is a person who knows something that others don't, or knows it better, or uses her skills more effectively. But you must also be acknowledged as such, by yourself and by others. Frequently, women are too modest and do not acknowledge their own expertise. We remain invisible, down-playing what we know, attributing it to luck. When you have specific *knowledge* or possess a unique skill, you also have a responsibility to share it with others, for you are then an invaluable resource.

Although *knowing* is a skill that can be sharpened, most of us already possess it in varying degrees. We use it almost unconsciously. To use it more effectively, we must acknowledge its force and use it consciously. *Knowing* is what provides the subtle clues upon which most of us base many of our decisions. We use this ability when we respond either favorably or unfavorably to a stranger. When you say, ''I just *feel* that this is wrong for me,'' you are in touch with your *knowing*. Men are usually skilled at attaining knowledge. They work at developing specific competencies and become *experts* at something. Women in general have good process skills: We recognize the significance of a single event and see the pattern in a series of events; we pick up nonverbal clues that others do not see or hear; we respond to what people are feeling, not just saying. This has given us the reputation of being intuitive.

Several of my women friends have told me of their feelings of connection with the universe, of being centered, of being in touch with their knowing ability on some deep, primal level. Some women seem to use the ability instinctively, without realizing it. However, when the ability is recognized, claimed, and cultivated, the results are usually even more significant. Yet many women, although realizing the power that knowing grants, do not acknowledge it. Because powerful women are often envied or feared, we stay under cover to avoid incurring sanctions. Effective power uses both skills. People at the top of the power hierarchy can deal with content and are task-oriented, yet they also know process and are people-oriented. Most men still need to develop their process skills; most women still need to acknowledge competencies in both areas.

However, it is important to realize that influence power is not necessarily the ultimate goal. You may choose to remain in a dependent power position. You may enjoy your job, like your co-workers, have a meaningful life outside of work and enough influence to get what you want done. You may not want more responsibilities, more work, or a greater tie to the organization.

Or your own path may lead you to an intermediate power position where you can be perfectly happy and content. At that level, you probably have a little more money and prestige. You most likely deal more with people than with tasks and do more planning than executing, yet you're not necessarily "in charge" and are therefore not burdened by that "charge."

I settled at the intermediate level myself when I decided to teach in a university. I share my one secretary with other faculty members and report to only one dean, whose support and advice I occasionally seek. There is no hierarchy except for faculty rank, yet I want to be well-regarded by my colleagues so that I can influence specific curriculum decisions and general educational objectives. I use my influence largely through process—that is, in the way I behave at faculty meetings; in what I comment on, and how effectively; in my committee participation; in the relationships I build outside the conference room. Of course, content is equally important—I must *know* what I'm talking about—but I will not hear others or *be heard* unless I pay attention to process.

The important thing is to find your own path to the power level that is right for you. Many of us don't know how to do that. Instead, we succumb to pressures from family and friends, letting them tell us to move up when we don't want to, or to stay put when we're ready to go on. We are often our own worst enemies simply because we don't know how to negotiate the roadblocks that are in the way. This book will deal with passing those roadblocks—both internal and external.

DEPENDENCE VERSUS ENTITLEMENT

There are two types of dependencies. One is the dependence upon tangible resources, such as money, staff, promotions, or equipment. The other is the dependence upon intangibles, such as being liked, respected, approved, or praised. The first dependence is external and is controlled by others; the second is internal and is controlled by ourselves. On our *Paths to Power* we will not travel from dependence to influence unless we can confront our dependence upon the positive regard of other people.

Although we each differ in the *amount* of positive regard we need in order to be happy, to feel esteemed, to work effectively, we all need to be recognized for who we are and what we do. The question is, How much recognition is necessary? How often should it be offered? By whom and in what way? When behavior is excessively controlled by the desire to please others or to gain approval, the demand for recognition is not realistic.

We need to learn to be less dependent upon the dispensation of goodwill. One way is to give it to each other; another is to understand its origins and its continued unconscious impact. Real freedom is knowing what needs to be done in order to prosper in the organization without having to rely upon the positive regard of other people in order to survive. To obtain that freedom we must become free of the guilt produced by transgressing the "shoulds" that were produced by socialization: We should *always* "produce, perform, have the right answers, be prepared, win, be a success, finish everything, be well informed, be consistent, be on time, be logical, be rational, impress others, trust authority, go through channels, be able to make it, be able to take it, and finally, be perfect."[1] When we achieve freedom from the "shoulds," we will attain freedom for action, freedom to make things happen, to be influential, to be powerful.

People with this freedom have something I call "entitlement," which is the belief that you have the *right* to be where you are, say what you say, do what you do, look the way you look, and sound the way you sound. You probably recognize it as a feeling of comfort with yourself, a feeling that you will not be unduly influenced by others, a feeling of being in charge. Entitlement is empowering. If you do not feel entitled to lead your life as you see fit, you will have a difficult time gaining influence power. Why is the feeling of entitlement so elusive? Why are we so easily intimidated? Why do we lack confidence in ourselves and our abilities? Some of it is due to the work environment, which prevents women from developing their strengths; some of it is due to the socialization women receive, a stereotyping process that starts at birth.

SOCIALIZATION

A recent study asked parents to describe their babies within twenty-four hours after birth.[2] Although hospital records showed no significant differences between the physical characteristics of boys and girls, the parents saw differences: Daughters were described consistently as much softer, finer featured, smaller, and less attentive than boys.

Another research project showed that, although parents *say* that both boys and girls should be treated alike, the parents' actions belie their words. In this study, eleven mothers, all of whom had young children of both sexes, were observed as they played with a six-month-old child in a nursery.[3] Five of them saw the baby dressed in blue pants and were told his name was Adam. The other six saw the *same* baby introduced as Beth in a pink dress. The mothers were told beforehand that the study concerned child-rearing practices, and all eleven were interviewed after the play period ended. Three toys—a fish, a doll, and a train—were placed on a table. The women who thought the baby was a girl handed over the doll more often. Those who thought the baby was a boy handed over the train. There was no difference in the handling of the fish. What is significant here is that these mothers were not aware of the different ways they treated the children.

The stereotyping begins at birth and continues with or without the awareness of the parents. Some parents follow socialization purposely. Others try not to yet are not aware that they give ambiguous signals. On one hand, they profess to have no different expectations for boys and girls; on the other, they do not act upon this. The child hears one message but is subjected to the attitudes and behaviors inherent to the opposite one.

No matter how aware a parent is, I believe it is next to impossible to totally free yourself of the prejudices and stereotypes inculcated by socialization. Once this fact is recognized, the consequences can be dealt with much better. What are those consequences? Girls get approval for "feminine behavior"; boys get approval for "masculine behavior." This approval begins at home and is further reinforced at school and by the hundreds of hours of TV that children watch.

In most households there still tend to be domains. For instance, in the kitchen "he helps her." But when she's chopping wood, "she's helping him." She says, "Thank you" to him for taking the children to the doctor; he says "Thank you" to her for taking the garbage out. The children see it all and know it all.

Even if the parents do share everything equally, differences, I venture, will be reflected in the answer to the following question: Who is the one who actually knows what is in the pantry and what items need to be bought? Usually, the woman is very subtly in charge. She is the one who notices where the dust is, when the children look feverish, and what needs to be bought, fixed, and cleaned. (It is not clear whether this "noticing" is due to the socialization process or to genetic factor. Recent research indicates that women are more sensitive to visual stimuli, to sounds, and to touch and are more aware of small changes.[4] This may be why we *notice* more and also therefore *react* to what we see.)

Little boys are encouraged to initiate, take risks, experiment, explore unknown territories, and defend themselves; little girls are encouraged to be friendly, quiet, obedient, and careful, and to appreciate protection from boys. What can we expect from these children as they grow up? Girls are taught to wait to be chosen, to be asked, and will feel lucky when this happens. On the other hand, boys will go, choose, explore, and decide, accepting all the concomitant risks. Girls will prefer to play with people they like; boys, with people who are skilled.

Popularity is one important measure of success for teenaged girls, and many spend most of their time trying to appeal to boys. This means not excelling at sports and sometimes not at studies. Since boys are taught early in life that girls are no competitive match for them, those girls who are somehow threaten a boy's masculinity. The price for excellence could be loneliness. For many girls, it is too high a price to pay.

If girls are to be protected, made fun of, humored, taken out, but not taken seriously, no wonder men have difficulty accepting women as colleagues. They have always seen them as mothers, sisters, daughters, teachers, nurses, secretaries—support roles. But there is some evidence that this strong socialization process is changing somewhat, as a result of the women's movement. Certainly there is more rhetoric about breaking down stereotypes. However, the most recent research seems to indicate that although many men *speak* in terms of equality of the sexes, they still *act* in discriminatory ways.

Changing the system

Even though awareness must precede action, they are very different processes, and what we are experiencing today is the time lag between the two. For modern women, this time lag between heightened awareness and the need for action presents a new problem. Many are experiencing a real gap between how they are *supposed* to feel and act and how they *actually* feel and act. All these discrepencies reinforce the inability of many women to identify the cause of their powerlessness. Should they blame themselves, their parents, the men, the organization, society?

Societal systems are made up of organizational structures composed of people. The people with decision-making powers can either open doors for those below them or keep the doors closed. To change the system, two things need to be done: The first is for us to learn the previously secret lock combinations on the doors so that we can open them ourselves. The second is to educate the men to make it easier for us and eventually for them, too.

A word of caution here about accepting the generalizations attributed by socialization to *all* women. Research shows much overlap between the sexes: women and men share more similarities than differences.[5] More and more women are being raised with masculine messages. Reinforcing the dysfunctionality of the socialization of *all* women may give a convenient excuse to the men who want to see women as inadequate.

Rosabeth Moss Kanter, at a 1976 UNH lecture, spoke of this sexual overlap as the *double bosom theory*. She suggested not to focus only on the differences between men and women, but also on their similarities (the cleavage).

The Double Bosom Theory

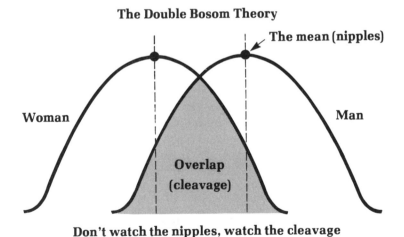

Don't watch the nipples, watch the cleavage

Reprinted by permission.

There are many things management should do to help women advance. However, this book is about what *we can* do, not what *they should* do, but keep in mind that often, if they won't, we can't. We can go just so far alone; after that point we need to unite with peers and support groups. Laws need to be enacted that will effect and enforce a change.

THE SEVEN A's

So here we are. Most of us are unwilling to tread the male *Paths to Power*, but we have no clear path of our own. Before we can forge one, we must remove the old stumps and underbrush, the internal roadblocks of our socialization, what I call the 7 A's: authority, assertiveness, accountability, accessibility, affiliation, approval, and affability.*

Authority

Authority, for most women, has always been in the hands of others, first parents, then teachers. This is normal for both growing boys and girls. Once in school, however, children find that status and authority shift to male figures. The books they read and the media they are exposed to all confirm this. The teachers are women, the principals, men; the nurses are women, the doctors, men. Police-"men," fire-"men," chair-"men." Men are in charge; women at best merely help.

   After all this conditioning, the adult woman may be uncomfortable in an authoritative role, in exercising her authority. I know many women managers who have difficulty asking their secretaries to do things that are perfectly legitimate. I know I often have trouble asking my secretary to retype a letter; it somehow seems punitive, and of course I want to be nice! No role models, no practice! Women must discover how to be and what to do, expecially since the male model of authority is not one many women want to emulate. I believe that the problem with authority is one of conviction. If you are convinced that what you ask of a secretary, a colleague, or even a boss is legitimate, if you know it will benefit you and the organization, (and a correctly typed letter reflects the standards of both you and your company), then you should be able to exert your authority with more ease. We feel powerless when we feel doubtful, not legitimate, unknowledgeable. This last can be a real trap—many women seldom feel they know enough or are good enough. Yet "good enough" will have to mean just that.

Assertiveness

The issue arises from role expectations. As a woman, you are in a double bind.† If you act as assertively as men do, you will be seen as aggressive, because men (and some women) are not accustomed to seeing women assert themselves. If you are sub-

---

*Adapted from Marcy Murningham, UNH lecture, 1977.

†The *double bind* is a psychoanalytical term that means you cannot do right by someone else whatever you do. The double bind is a no-win situation.

missive, compliant, and dependent, they will assume that you lack leadership qualities and, therefore, managerial potential. Whichever way you go, you might encounter anger, resistance, discounting behavior, or indifference. Do you feel stuck?
Address this issue! When you confront the men, you may be seen either as paranoid
or as making mountains out of molehills, but you *must* do it. When enough women
consistently act with assertion, when enough men consistently become more aware,
role expectations will begin to change.

Accountability

Accountability is the responsibility that we, as women, carry with us. We are expected to be harmonizers, mediators, the ones who understand. If there is a conflict,
we are expected to smooth it over. If the children are misbehaving, at home or at
school, it is somehow our fault. If the house is not clean or the meals are not on time,
we are accountable. If the laundry is piled up or the floors need waxing, we are supposed to do it. If our parents (sometimes even our husband's parents) are sick, we are
expected to take care of them. And even if a husband or friend shares the housework,
we still feel responsible for the relationship, the home, the children, the marriage,
the family. Whatever a man does at home, he only "helps"; it is we who are ultimately accountable.

As managers, we may have trouble delegating responsibility and instead feel that
we ought to do it all ourselves. When we *do* delegate, we may feel guilty or worry
about whether the work will be done properly—we might supervise too closely. We
have to learn to unburden ourselves and to trust others more.

What might help is to ask ourselves, "Am I *really* accountable for this? What *are*
my responsibilities?" Perhaps someone else can share them. If we can let go of the
feeling that "it won't be done well enough, fast enough, or in the way I would do
it," we will be freer to pursue more rewarding tasks. If our way is always the only
right way—if we are perfectionists—leaving the lowest rungs of the hierarchy will be
very difficult. We are so good at staying there.

Accessibility

The issue here is in the difficulty that most women have in setting boundaries.
Women are so accustomed to meeting the needs of *other* people—husbands, children, family, friends. Society has always said that this was woman's most fulfilling
role. But what about our own needs? Do we fulfill them by meeting everyone else's
needs, or do we neglect them by doing so? Often, the boundaries are fuzzy and hard
to define.

This dilemma becomes apparent at work by the tendency to be available to anyone who wants to come and talk. We listen, try to help, give some time. We cannot
cut salesmen off, or refuse to pitch in when our help is solicited. We are more available than men to employees, to students, to clients, or to patients. We must learn to
close our doors and say, "No, I'm too busy now," so we also can pursue other tasks.

I just finished a research study in which I interviewed 102 female managers and
68 male managers on dimensions of accessibility. The most significant finding was

that women are *twice* as accessible as men in similar positions. Not only did they make themselves available to employees, colleagues, and clients, but they left their offices more often to *go* and *see* if they could be "helpful."

This may indicate a fear that unavailability might be construed as rejection. Or it may be real caring about employees. But either way, it is important to shut yourself off once in a while in order to think quietly and creatively, to work without interruptions. This may have to take priority over being available to discuss every detail.

Affiliation

The problem lies in feelings of dependence on existing friendships. Friendships are very important to women, who early in life develop the prerequisite of liking other children in order to play with them. Not so for boys. Boys play for the pleasure of the sport; girls play for the pleasure of being with each other. This is not to say that boys don't have friends, but their friends serve a different purpose. For girls, a friendship is formed to share intimate thoughts and feelings; for boys it is formed to share experiences.

Translated on a managerial level, women feel loyalty to their colleagues and peers. It is, therefore, more difficult for women to change from one department to another, to move on, to move out and leave their colleagues behind. This naturally has an effect on career mobility. What should our priorities be? Up the managerial ladder and away from our friends? Because we are breaking new ground, we need the support of friends; it is difficult to go it alone. But if we also don't want to remain stuck where we are, the dilemma can be paralyzing. It is critical for women to seek new support groups, preferably at the higher levels of the organization.

Upward mobility is not the only way to go. You may prefer horizontal enrichment. Pay attention to the influence of family: "Of course you'll take that new position," and "You must think of your future," or of peer pressure: "Stay with us, we're your friends," and "You'll be alienated up there." You must decide what is right for you.

But harder than leaving your former colleagues as you move up and away from them is delegating tasks to them as they become your subordinates. It's very hard to be demanding, exacting, and even critical of people you consider your friends.

Approval

Because many women do not think of their jobs as part of a career path, the rewards for work well done must come from someone's recognition of that fact. Therefore, women seek approval rather than constructive criticism. Any disapproval is then considered to be directed at the total person rather than at only a single action. In other words, women tend to personalize criticism. If I am told I did not teach well today, I don't hear that today from 10:00 to 12:00 I did not conduct my class well on a particular subject. What I hear is that I am a bad teacher, from head to toe and forever! This is a little exaggerated, but at the moment I am faced with the criticism, that is exactly how I feel. Women need to care less about what others think. It's important to differentiate between what is thought of us as people and what is thought of one piece of work or of one action. Besides, who is doing the evaluating? Is that person your judge? Have we attributed power to that person because we *need* approval? We must learn to listen to criticism and see if there is anything to be learned from it. If not, it should be discarded. What helps me do this is to remember that when Peter gives feedback to Paul, he is saying more about Peter than he is about Paul.

Affability

Women are not supposed to muddy the waters, to stir up conflict; instead, we are supposed to smile, be nice and polite, be gentle, be calm, be mild, be helpful. It's never "their problem," it's always "ours." How, then, can we confront or criticize?

Orientals are considered enigmatic because they smile a lot. Many men find women equally difficult to read because they, too, smile a lot. We need to stop smiling and tell it like it is! We must stop discounting ourselves, stop qualifying our statements, and stop worrying about everyone else's ego! We pay a big price for being nice often when we don't feel nice at all. We need the *freedom* to express our feelings without being sanctioned for "unladylike behavior." The only way to do it is to do it.

THE FIRST STEP

The *Paths to Power,* to self-confidence, to self-realization may be many, but they all have one starting point: awareness. We need awareness of ourselves, our needs, our rights, and our aspirations. This awareness does not come full blown into our consciousness, but develops slowly, a step at a time, with frequent regressions and several full stops. Some of the 7 A's will remain with us all our lives, but in milder or at least more controllable forms. I do not need as much frequent *approval* from others as I used to, and I feel perhaps just a bit less *accountable* for everything. I made great strides in *accessibility* when I finally learned to say "No" to protect my privacy and my time. *Assertion* used to be an issue for me, but I do assert myself more as I gain in self-confidence. I am fairly comfortable when in *authority,* although I become least authoritative then, but I struggle for some form of influence when in a powerless position. *Affiliation* will probably remain a strong need all my life, and I am still more *affable* than I wish I were.

How about you? Which of the 7 A's give you the most trouble? They may not be experienced by everyone in the same way. After you become more aware of these potential barriers to your power, you have two alternatives. One is to sigh and say, "I can't help it, that's the way I'll always be." This is the powerless stance, the dependent position, a victim behavior. The second is to say, "There is not much I can do about the way I was brought up, but I do not *choose* to repeat the dysfunctional patterns." This is empowering yourself.

The first step in being powerful is to *feel* it. Stand in front of a mirror the way you usually stand, and then put your feet wider apart. Throw your shoulders back an inch, move your head towards the ceiling, look yourself in the eye, and say, "I'm entitled." Say it again, loudly, clearly—with assurance! For you *are* entitled. The mere fact that you are reading this book means you are curious, you are ambitious, you want to learn, you have confidence that you can make it.

You may not have any control over how bright you are, how attractive you look, who your parents were, or how you were socialized, but you *do* have control over how hard you are planning to work, how motivated you are, how much you are willing to learn, how many risks you are willing to take. These first tentative steps of awareness have consequences: You start to behave in different ways, make new demands, expect better treatment, make waves. You can expect that many men—and some women—will react to these changes. Here is a chart that indicates how some people may react toward your new stages of awareness. Although not everyone will respond exactly this way, it will give you some idea of what to expect.

Women's power through awareness*	Men's reactions
Happy to be dependent. Denial of women's issues.	Collusion in denial of women's issues.
Discomfort at dependence; mixed curiosity and envy of professional women. Beginning identification with women's issues.	Identification with the individual women who have issues, alternately denying and sympathizing.
Anger at own powerlessness and dependence. Target: individual men and/or society.	Backlash of anger as individuals; confusion as to societal guilt.
Depression at futility and slowness of change.	Individual guilt as perpetrator of discrimination; frustration with having to deal with it.
Action as an individual to become assertive.	Upset at new demands but sometimes admiring the courage.
Action on a group level; excitement at opportunities.	Threatened by group action and by potential loss of opportunities.
Taking power.	Fighting back.
Sharing power.	Relief and acceptance of new colleagueship, or entrenchment into positions of denial.

*Adapted from Judith Palmer, "Stages of Women's Awareness," *Social Change* 9-1, 1979.

Although men may react unfavorably to the newly found power of women, in the long run it's to everyone's benefit. To think in terms of masculine and feminine types of behaviors polarizes the sexes. Therefore, neither men nor women can broaden their range and be whatever feels right for them for the particular situation. Do masculine and feminine traits *have* to be at odds? Is being soft the opposite of being tough, being passive the opposite of being dominant, initiating the opposite of following, listening the opposite of talking, praising the opposite of criticizing? I don't think so—I think you can be masculine *and* feminine at the same time.

Women of today—and that means us—are trail blazers. How can we get from here to wherever it is we want to be? First, we should decide if the grass ''there'' is really greener; we may find that the bluebird has been singing in our own backyard all along. Second, we must set the goal: the level of power we want. Without a goal to work toward, we will not get there. Third, we must get in touch with our knowing and expand our knowledge of the world of work.

We can remain on the outside, refusing to play because we don't like the rules, but remember: Those who don't play don't usually get to make or change the rules. Only the insiders, the players, have a chance to modify them. This book will help women learn the rules. We *can* follow the *Paths to Power* without losing ourselves on the way. We *can* be powerful and keep our integrity, our connections to our nature, and our relationships to other women. We *can* influence others and still be sensitive, caring, and warm.

NOTES

1. Adapted from Theodora Wells, ''Handling Sexism at Work: Non-Defensive Communication,'' in *New Life Options: The Working Woman's Resource Book*, ed. by Rosalind K. Loring and Herbert A. Otto (New York: McGraw-Hill, 1976), p. 130.

2. Jeff Rubin, Frank Probenzano and Zella Luria, ''The Eye of the Beholder: Parents' Views on Sex of Newborn,'' *American Journal of Orthopsychiatry* 44-4 (1974).

3. Jerrie Will, Patricia Self, and Nancy Datan, paper presented at the American Psychological Association Meeting, 1974.

4. Daniel Goleman, ''Special Abilities of the Sexes: Do They Begin in the Brain?'' *Psychology Today*, November 1978.

5. David Tresemer, ''Assumptions Made about Gender Roles,'' in *Another Voice: Feminine Perspectives on Social Life and Social Sciences*, ed. by Marcia Millman and Rosabeth Moss Kanter (New York: Doubleday, 1975), pp. 308–339.

SELECTED READINGS

Bunker, Barbara Benedict, and Seashore, Edith W. ''Power, Collusion, Intimacy, Sexuality, Support: Breaking the Sex Role Stereotypes in Social and Organizational Settings.'' In *Beyond Sex Roles*, ed. by Alice G. Sargent. St. Paul, MN: West Publishing Co., 1977.

French, J. R. P., Jr., and Raven, B. ''The Bases of Social Power.'' In *Group Dynamics: Research and Theory*, ed. by D. Cartwright and A. Zander. New York: Harper & Row, 1960.

Harrison, Roger. "Understanding Your Organization's Character." *Harvard Business Review,* May–June 1972.

Jourard, Sidney M. "Some Lethal Aspects of the Male Role." In *The Transparent Self.* New York: Van Nostrand Reinhold, 1964.

Korda, Michael. *Power: How to Get It and How to Use It.* New York: Simon & Schuster, 1977.

Kotter, John P. *Power in Management.* New York: AMACOM, 1979.

McClelland, David C. *Power, the Inner Experience.* New York: John Wiley & Sons, 1975.

Mechanic, David. "Sources of Power in Lower Participants in Complex Organizations." *Administrative Science Quarterly* 7–3(December 1962):349–364.

Stein, Barry A., and Kanter, Rosabeth Moss. "Management Psychology." *Board Room Reports,* June 30, 1977.

Varoff, J. "Process vs. Impact in Men's and Women's Achievement Motivation." *Psychology of Women* 1–3(Spring 1977):283–292.

Chapter 2

SHE WHO GETS HIRED

**She who gets hired
Is not necessarily the one
Who can do that job best
But the one who knows
The most about
How to get hired.**

The Preparation

Like any journey, the one that starts you on your path to power must begin with preparation. Good planning can make all the difference between a successful trip and a disaster. If you forget your raincoat you might get soaked; if you forget your bathing suit you might not have a good time. You need to learn some of the language and know what to expect from the natives.

On your journey you'll need to take with you:

- a helmet for the knocks;
- a cushion for the falls;
- a mop for the tears;
- earplugs for the gossip;
- good shoes for running twice as fast as the others in order to get to the same place at the same time;
- a hammer to nail down promises;
- a key to open closed minds;
- a hatchet to open closed doors;
- a gavel to command attention;
- a microphone so that you'll be heard;
- a box to pick up the pieces;
- and a friend for the good times and especially for the bad times.

And you might as well pack:

- ◼ a certificate of merit;
- ◼ a gold star;
- ◼ the medal of honor;
- ◼ the purple heart;
- ◼ a badge of courage;
- ◼ and a halo

for when you have arrived.

As with all travelers to new lands, you'll have doubts, anxieties, excitement, and many, many questions: Where should I go? How far? How soon? Whom will I meet? What will happen to those I leave behind? Will I return? Will I change?

WHAT IS IN YOU? WHAT IS OUT THERE?

Whether you're just out of school or entering the job market after raising a family you need to think about three things: (1) what is in you, (2) what is out there, and (3) how the two will fit.

Let us start with the first. Who are you? What is *in* you? Write down how far you want to go. In other words, how high up is the top for you? Do you want to be president? Would you rather supervise a few people? Or do you prefer to work alone?

A word of caution here: Moving up is not the only way to go. Many people very wisely do not yield to societal pressure to go up the management ladder. They know themselves well and instead look for enrichment at their present job level. If you are content where you are, there is no reason to move on. If you feel satisfied and challenged and have good relationships with your co-workers, your fulfillment needs may have been met. It may be that you have a full life outside of work and do not wish to put any more energy into competing, learning new skills, or playing politics, which you need to do to get to the next rung.

Challenge can also be found in horizontal moves. You may want a different job but at the same level as your present position—or you may add more responsibilities to your job. Upon typing this draft, my secretary Jenifer McKinnon said, "Why don't you write about the perfectly happy secretary who just wants to stay on in her job forever?" So I just did, but in the meantime Jenifer typed another five chapters, which, she said, gave her the courage to take a real estate exam. I lost a good secretary, but she is a good real estate agent, loving her work and doing extremely well.

Now write down *why* you want to move up, if that's what you decided. What will it do for you in terms of feelings about yourself? In what way is it meaningful to you and significant to others? Are there any prices to pay? By whom?

What it is that you *like* to do and what is it that you *can* do? How do you see your preferences and your competencies fitting together? What do you need to learn to be able to do what you want to do? Where can you obtain that knowledge or that skill?

What are the obstacles that stand in your way? Which people can help you overcome those obstacles?

Think about *where* you'd like to work. You may like certain parts of the country, such as the East Coast or the Middle West, either because it is what's familiar to you and you already have established friendships, or because you want to go to a new place. Do you prefer mountains or the seacoast, a warm or cold climate, an urban or rural setting?

Give some thought also to the *kind* of organization you'd like to work for. There are a couple of issues to consider. First, would you prefer working for a large company or a small one? Some of the answers may be found within your own experience; for instance, whether you enjoyed being in a small school or a big one, whether you are more comfortable with anonymity or intimacy. Contrary to expectations, you can find friendship working in a close-knit unit of a large company. The difference in opportunity is not clear between the two alternatives. Even though at first glance you might expect a large company to have more opportunities, it is also possible to get stuck in the lower rungs forever. In a small company you are more visible, and if you do well, perhaps you'll have a chance to move on, unless all the upper echelon slots are taken and there simply is no room at the top.

Do you want to work for a manufacturing company or for a service organization? Although there is usually more money to be earned in industry, service and nonprofit organizations tend to have a family atmosphere and are often run like a family business. It is my experience, and that of some of my colleagues, that schools, hospitals, and social agencies foster closer interpersonal relationships than does industry. Of course, this intimacy also breeds the bickering to be found in any family situation.

Think of what kind of work atmosphere you'd like. Would you prefer a traditional organization, very formal and structured, such as a bank or insurance company? Or would you feel more comfortable in an innovative organization, such as an advertising agency or a firm on the edge of new electronics technology? In these situations, creativity is rewarded but the anxiety level is higher. Decide whether you enjoy the challenge of being a pathfinder or whether you would rather leave those hassles to others and follow on some already-beaten track. What is risk taking for one person is no risk at all for another.

Another major question you should ask yourself is, "What can I do that is marketable?" The other side of that coin is, "Who would hire me?" If you don't think your liberal arts college degree is enough to land you a job, or if you have been out of the work force for so long that you don't know where to start, take heart. You have all kinds of skills that will be valuable to an employer. You just need to figure out exactly what they are and how to present them. You need to translate your competencies into credentials because that employer out there is going to look at a piece of paper—your résumé—that certifies you're good enough to be hired.

Credentials and competencies

What are credentials? The dictionary says: "That which entitles one to confidence, credit, or authority. Written evidence of qualifications." But let us take a look at the definition of competencies: "The state or quality of being capable or competent;

skill; ability.'' So all you need to do is transform what you already know and have done into a ''written evidence of qualification'' that will entitle you to ''confidence,'' yours and others', and to ''credit and authority.''

The purpose of the following list is to help you translate some of the things you do or know into ''marketable'' skills. The italicized words are the credentials, which are your competencies translated into managerial terms.

For instance, if you ever taught someone to ride a bike, you can *instruct*.

If you ever organized a car pool, you can *coordinate*.

If you ever had to figure out the cause of a plugged-up drainpipe, you used your *analytical* abilities.

Managerial function	Student equivalent
Planning	Determining what major will best fulfill your career objectives.
Staffing	Choosing someone to type a paper for you or to take notes when you're absent.
Representing	Being a student representative; speaking to an instructor on behalf of other students.
Negotiating	Getting a better dorm room or an apartment off campus; deciding with roommates who does what chores.
Mediating	Dealing with conflict among family members; confronting differences between yourself and others; harmonizing.
Organizing	Setting priorities between the various demands of school; managing your time between work and play.
Directing	As a student group leader or a team leader, instructing members as to your expectations and standards of performance.
Controlling	Changing your study habits after getting a poor grade.

If you ever saved money for a trip, you had to *estimate* the cost of it.

If you ever budgeted your expenses, you had to know how to *forecast*.

The phrase ''People with experience wanted'' looks like a barrier to the world of work. Most of us are used to looking at only traditional paid employment positions as the type of experience that is wanted, but we learn competencies from a variety of experiences. If you've never worked in a paid position, you may be thinking that you would never qualify for a management job. But whether you're a homemaker or a student, you've no doubt performed managerial functions literally dozens of times. The tables on these two pages show some typical examples.

Managerial function	Homemaker equivalent
Planning	Determining the household income and budgeting accordingly (food, clothing, rent, insurance, transportation, recreation, etc.).
Staffing	Choosing a babysitter you can trust; hiring dependable live-in help.
Representing	Being a spokesperson on behalf of your children.
Negotiating	Contracting with service people; getting a better deal on a car or an appliance.
Mediating	Dealing with conflict among family members; confronting differences between yourself and others; harmonizing.
Organizing	Deciding upon the maintenance needs of your household; determining which tasks you can do and which you will delegate; determining the order of the tasks.
Directing	Instructing family members or outside repairpeople on what needs to be done, when it needs to be done, and your standards of performance.
Controlling	Informing a plumber that the contracted work is unacceptable.

MAKING THE FIT

You need to learn to translate your skills into terms that fit the competencies required in a specific job. Let's take as an example a woman who has organized a car pool to either get to work herself or to take her child to school. The specific skills refer to what she *needed to do* in order to accomplish her purpose. The general competencies refer to what she *had to know* in order to do it. They can be applied to other experiences. As you will see, skills and competencies become her credentials for a job.

Would you say that the woman who organized a car pool could supervise a large typing pool and be responsible for the flow of paperwork in an organization? If you look at the specific skills and the general competencies necessary for both, you would see a good fit.

We all have experiences that can be analyzed in terms of specific skills and general competencies. If you're still in school, perhaps an analysis of your participation in an extracurricular activity or sports team will reveal skills and competencies you were previously not able to articulate. A summer job is another likely candidate for analysis. In fact, many experiences can be analyzed to see if any patterns emerge. Here is where a friend or support person can be particularly helpful. He or she will be able to help you remember in more detail just what you did, and what you had to know in order to be able to do those things. He or she also may counteract that part

Organizing a Car Pool

Actions	Specific skills	General competencies
Asked around to find out who might be interested in car pooling.	Articulating goals; recruiting.	Initiative; communication skills.
Found four people who were willing to do it.	Motivating people.	Ability to influence others.
Decided with them the route to be followed for pickups.	Identifying needs.	Decision-making ability; logic; listening skills.
Discussed how to deal with last-minute changes (illness).	Anticipating problems.	Negotiation skills.
Checked that drivers are safe and dependable.	Evaluating performance.	Synthesis of information; quality control.
Talked to an often late person about being on time.	Confronting.	Interpersonal skills; tact and diplomacy.

Organizing a Typing Pool

Actions	Specific skills	General competencies
Get to know staff's abilities and limitations and establish team spirit.	Motivating people.	Ability to influence people.
Establish the priority of the work to be done.	Identifying needs.	Synthesis of information; decision-making ability; logic; listening skills.
Distribute work equitably.	Facilitating the task.	Decision making; leadership
Make a contingency plan (in case of absence).	Anticipating problems.	Negotiation skills.
Check for quality of output and timeliness.	Evaluating performance.	Quality control.
Talk to people performing below standard.	Confronting.	Interpersonal skills; tact and diplomacy.

of you that may not be comfortable admitting that you can do wonderful things. This process helps to focus on answering the question, "What can I do?"

There are multiple benefits in going through a process such as this. It is empowering, enabling you to think of your abilities in a more confident and realistic manner. It forms the basis of your own marketing effort toward finding a job. The traits you list could be carried from job to job and are applicable to any position. Listing the traits on paper is effective and makes a positive, assertive impression. It also has a longer-lasting impact than a verbal statement. This process will help to identify areas that may need improvement.

It is not enough, however, to just identify your competencies. The next step is to persuade the employer that your competencies are useful to the organization. The way to do this is to look at the competencies needed to perform a particular job effectively. Employers do not often know how to break down a job description into special skills and general competencies, so you'll have to do some of that work for them and possibly teach them to think in those terms. You have to learn to ask the right questions—this is critical for women entering the job market. You'll be seen as quite impressive, for it will prove your ability to conceptualize, to form ideas and think in terms of analogies.

The five fundamentals

Besides the appropriate technical skills needed in specific professions, women also need five fundamental skills to survive in the world of work.

1. *Personal skills,* also known as intrapsychic skills.   These refer to our awareness of ourselves, for example, how we react to criticism or how we feel about praise. Some of us may look at feedback defensively, saying "It's *their* problem"; others may accept it indiscriminantly, saying "It's all *my* problem." Some will welcome it, others will feel judged. Personal skills refer to seeing ourselves as both actors and observers; the observing part of us looks at the acting/feeling/thinking part of us in an attempt to detect patterns; to ascertain whether our reactions are functional or dysfunctional, appropriate or inappropriate; to understand the impact others have on us.

2. *Interpersonal skills.*   These are the skills at work in any two-person relationship. Interpersonal skills refer not only to mutual understanding, but also to the interplay between two individuals. When a colleague of mine is seen as very easy to get along with and I too am very easy going, but the two of us do not get along at all, there may be nothing wrong with either of us, but something is very wrong in our interactions. The interpersonal skills would involve understanding what buttons you push in the other (and vice versa) and being able to act differently according to that understanding. Good interpersonal skills would entail being able to deal effectively with all types of personalities in a variety of situations, and understanding the impact we may have on others.

3. *Group skills.*   We use these skills as participants in such groups as work units, project teams, task groups, office parties, support groups, and professional meetings. As team members we can be effective or disruptive, helpful or unnoticed, supportive or confrontive. A person skilled in group dynamics understands what is going on and knows how and when to intervene most effectively to enhance the work of the group. Group leaders, whether formally or informally appointed, must understand whether to push for completion of the task or to pay attention to the people involved.

4. *Intergroup skills.*   Whenever there is more than one group in a work place, there are intergroup dynamics that influence people's behavior. Understanding the interplay between collaboration and competition is important for both the group members, the group leaders, and the managers who may be outside the groups. It is only through the understanding of members' attitudes towards members of other groups or towards other organizations that we can influence the course of these groups' relationships to each other. This is an important skill for any member of an organization, but the higher up we go, the more necessary this understanding becomes.

5. *Organizational skills.*   Women who are ambitious and wish to move up need to understand the structure of the organization in which they work. They need to know the formal relationships as well as the informal ones where often the real power resides. Each organization has its own path to power and this path can be identified and staked out. Organizational skills include thinking in terms of interrelated systems and understanding your present place and your future opportunities within your company.

Organizational structure

Let us look at basic organizational structure. Most businesses are divided in terms of functional areas. The first is the *financial area,* which deals with the acquisition and utilization of capital and with the distribution of profit. The second area is *marketing,* which deals with product selection, pricing, and promotion, and the channels of distribution. The third area, *production,* deals with product policies, such as "making or buying"; with facility policies, such as location and capacity; and with the selection of vendors (the sales force). The fourth area is that of *personnel,* which deals with staffing, development, compensation, and labor-management relations.

The *financial manager* oversees accountants who record all the company's transactions in order to give a clear picture of its operations and financial situation. The budgeting department decides on wages, purchases supplies and equipment, and estimates the profit from expenses and income. Financial forecasting determines if and when new product lines are developed, whether new employees are hired or a cut-back is ordered, and how the company's liquid assets should be invested.

The *marketing manager* sees to it that the company's merchandise or services flow smoothly to the largest possible body of consumers. This involves market research (determining the consumer's wants and needs), pricing (determining how much the consumers are willing to pay), promotion and advertising (determing how to persuade people to become consumers), and distribution (determining how best to deliver the product or service to the consumer).

The *production manager* must make sure that the company turns out quality products, in the right quantity, within the specified time, and with the lowest cost.

The *personnel manager* is in charge of human resources, instituting company standards for labor relations, employee benefits, motivation and morale, and job enrichment.

Now that you're armed with some ways of describing what you're interested in, you need to find out more about where that kind of work is done, what kind of people do it, and what the organizations are like that hire such people. How? The best way to get information is through personal contacts. Off hand, you may not think that you know anyone who can help you, but *please* don't stop your personal contact search prematurely. Ask everyone you know if they know someone who knows about the field you're investigating. As you start to gather information, you'll be able to be more specific about your choice of work, and you'll be able to identify some actual organizations that might hire you.

THE JOB HUNT

The job-hunt process is a long one. Some have said that it takes as long to land a good job as it takes to have a baby. Be prepared for months of waiting and hunting, and waiting and seeking, and waiting. . . and a lot of ups and downs. Some days you are going to feel terrific—excited about all the things you've discovered you can do, about getting ready for an interview, about hearing positive feedback on the way you are coming across. And some days you may feel just awful—you can't do *anything,* or you'll never find a good job, or you'll never get your foot in the door anywhere. All I can say is live with those ups and downs, and when you're feeling really discouraged, try to remember that it is a temporary feeling.

There is *excitement* at having made the decision to seek a particular kind of job that includes a flurry of activity—talking with people, getting names and addresses, polishing up résumés. Then there is the initial *rejection* stage, not receiving an instant "perfect job for you" offer within a couple of weeks after contacting potential employers. The first reaction here is depression, "I'll *never* get a job!" Fortunately, you move into a third stage: *realistic self-appraisal*, asking yourself how to improve your job-hunting technique. Another attempt should be made to find more and perhaps more suitable people to contact. Then you may perhaps meet with more rejection and depressions, and another stage of *redetermination* until you have eventual success.

An important point to remember is that we tend to be more attractive to employers and to others who might help us when we feel good about ourselves. The more confident we are, the easier it is to call Manager A and ask to schedule an interview. I know, for instance, that my voice doesn't quake when I'm feeling self-confident. So when I'm discouraged, I try to turn to someone who can help rebuild my confidence. Ask friends to review your letters of inquiry and to practice interviewing with you. The more support you can get as your search continues, the more easily you may be able to bounce back from those times of discouragement.

There are several specific steps to take in your job hunt. Each one will move you closer to finding a job that is right for you.

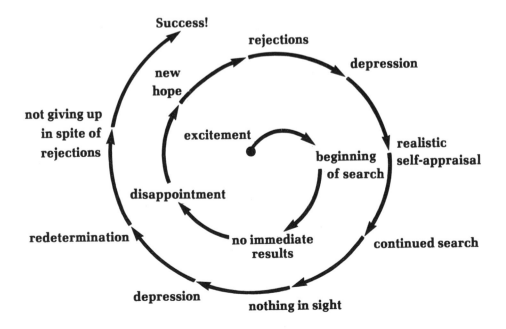

Step one: Prepare a good résumé

You have identified what is in you that seems to fit with what is out there. Now you need to make that connection evident to others. You do it through a résumé. Résumé writing is not easy, but it may well be one of the more important pieces of writing you'll ever do. Your résumé is your credential. It should identify competencies you acquired doing volunteer work, or competencies you own through your personal attributes. A résumé has two parts. The first part, the core, explains who you are and how you wish to be seen; the second part is directed to some specific person and to the purposes and objectives of some particular job that you aspire to fulfill. The core would list general competencies and the more important special skills you have; the second part would include only that experience relevant to the job for which you are applying. For example, do not write that you type if you do not want to use your typing skills in your next job.

Most résumés include the following information:

1. Name and address

2. Professional or career objective (optional, but important—it tells the employer your goals)

3. Employment experiences or work history

4. Educational background

5. Personal information (optional; list languages or travel if they relate to the job)

6. Interests list (optional; only if they relate)

7. Personal data (optional; use to your advantage only)

8. References (usually stated as being available upon request)

How you arrange this information depends on what type of résumé you choose to write. See pp. 237–248 of Appendix A for the different kinds of résumés, and choose the one that best fits your background and experience.

Perhaps one of the more important skills women need for résumé writing is an ability to translate their duties into terms that signify responsibility, using words that show active ability and power. Start your sentences with "I was in charge of," "I was responsible for," "I created," "I developed," "I ran," "I managed," "I led" rather than "I helped with" or "I had experience with." Write: "Because of my efforts, such and such happened." If anyone ever asked you to speak anywhere, mention it: "I was invited to discuss. . . ." It is a good idea to include your personal reasons for wanting the job, plus the professional reasons as to why the company would hire you. These two items are part of the covering letter, which is sent along with the résumé (see Appendix B, pp. 250–252.)

It has been traditionally difficult for women to put their best foot forward and to speak well of themselves. Bragging is not ladylike. Résumé writing treads a fine line between tooting your own horn and being justifiably proud of your accomplishments. The only way to know if you are doing it well is to have several people read your résumé and help you with it. The ultimate test is to imagine yourself as the employer: If you were that person, would you hire yourself on the basis of your résumé? Use the résumé critique form on page 249 in the appendix to answer this question.

There is an issue in résumé writing that concerns women alone: personal matters. Legally, you do not have to mention family. However, if a man writes that he has a family, he is seen as stable; if you do, you may be seen as having too much responsibility outside of work. Men don't have to contend with these prejudices. If you are willing to relocate, I suggest putting this fact in your résumé. There is a general assumption that women are less mobile than men, so if you are willing to travel, say so.

The problem for the older woman who has stopped working for any period of time or has never worked at all is the gap in her résumé for the years spent in child rearing. It is important to mention the skills required to perform any volunteer activities that were undertaken during that period. However, if you did not do so, write honestly that these years were spent in family responsibilities, listing some of them. Often a woman who has not worked steadily outside the home will feel that her résumé is inadequate, that she has nothing to show. This is where the whole issue of competence comes into play. I cannot stress enough that women *do* possess work-related skills—they simply discount them.

The web and woof of women's lives We have had to *plan* our day and our children's day, and juggle car pools, bus schedules, lessons, and dental appointments. This is the web and the woof of our lives. We have to *supervise* our children, *talk* to their teachers, *persuade* school principals, *discuss* prices with salespeople, *argue* with store owners, *check* on the quality of merchandise, *see* if it fits our needs, *negotiate* with plumbers and electricians, *budget, entertain*. We *motivate* our children to do their lessons, *negotiate* with our husbands about job sharing, *create* an atmosphere conducive to work or play, *comply with or refuse* requests, and make *demands*. Every time you tell the butcher that yesterday's cut of meat was excellent, you motivate him to keep up the expected performance. Every time you complain, you set expectations as to your standards.

We have always done it and continue to do it all the time. The difficulty is in translating all of these skills we already have to a different setting—the office. We are already competent administrators, all we need is practice in a new setting. Please don't discount yourself because you have not worked in a job without interruption. Get in touch with your accumulated experiences, your acquired wisdom. Identify it for yourself and then share it with others.

Step two: Develop a plan of attack

Your chances of finding a good job depend a lot on how you go about doing it. A haphazard search rarely produces the kind of oppportunity you want. Finding a job takes concentration, discipline, persistence, and most important, a plan. Write out your plan, then stick to it. This will increase the possibility of finding the best job available for you.

First day in the life of a future working woman:

8:00 A.M.   Look over newspaper ads; the most useful are the trade papers. Read *The Wall Street Journal* and *Business Week.* (Note: Only 10 percent to 40 percent of all jobs are found through advertisements; all others are obtained through personal contacts.)

9:00 A.M.   Call up employers and employment agencies (but only those recommended) who advertised. Schedule interviews.

10:00 A.M.   Call up local schools, churches, the YWCA, to find out if career workshops are given.

11:00 A.M.   Call up feminist organizations for leads.

12:00 Noon   Have lunch with a sympathetic friend who might know someone who might know of a job opening or can refer you to a company official to discuss job possibilities; set up appointments.

1:00 P.M.   Go to the local library and consult librarian for references on the various companies in the area you might be interested in.

3:00 P.M.   Buy conservative suit or a blouse to go with one you already own; buy a briefcase.

5:00 P.M.   Drink hot cup of tea at home or have a drink with a friend who could be helpful.

7:00 P.M.   Decompress; enjoy a relaxed evening.

Next day, follow up on all leads.

Once you have identified several companies that seem to be good places to work, make a plan for getting an interview and eventually a job.

Try to find someone who knows a company you're interested in. Try to get as much information as possible about the company—what its specialty is, what kind of people they look for, what it takes to succeed there.

Write a letter to the manager of the department in which you'd like to work, mentioning the person who recommended you. Where possible, avoid writing to personnel departments. If you can't find anyone who knows someone in the company, try to find the names of people who work there through sources like *Standard & Poor,* a financial reference found in your public library.

In the letter, describe something about yourself that fits the organization. It doesn't have to be a traditional employment experience, it could be a hobby that fits with the company product, such as tennis with the AMF Company.* Include your résumé and mention that you will be in the area shortly and will call to see if you can set up an appointment to talk with either the person to whom you are writing, or with someone that he or she might suggest about possible opportunities for you in that company.

*American Machine Foundry, Inc.

Call as planned. When the secretary answers, ask "Is _____ in?" If not, say something like "I sent a letter a few days ago, and I'm following up on that letter. When would be a good time to reach _____?" When you do get through say, "Recently, I wrote you a letter expressing an interest in _____[career choice]_____ at _____[name of company]_____." If the manager does not remember the letter, be prepared to summarize its contents. Then ask for an appointment: "I will be in your vicinity the week of _____. Might I make an appointment with you?"

Often the person will say they would love to talk to you, but there are no immediate openings at the company. You may want to make the trip anyway, because that person will probably refer you to others if you hit it off during the interview, and something may come up in a while. Besides, if something does come up, you'll be chosen over someone they have to interview from scratch—you've already shown an interest in the organization and presented yourself well during an initial conversation! Remember to always send a thank-you note after each interview.

If you're really interested in that particular company, find reasons to keep in touch. Look for articles about the company and forward them, along with a short note saying something like "Thought you'd be interested." Remind them of who you are; if you move, send them your new address. That way, you stay fresh in their minds, and if something does come up, you'll be one of the first they consider for the position.*

In general, it's worth pursuing almost any lead that sounds reasonable. The more places you send your résumé, the better your chances of finding a good job. To keep track of your plan, use the following job-hunt checklist.

Several of my students received job offers while interviewing managers as a class assignment. Try it yourself. Interview managers for a personal project, such as an article you're planning to write (some day) on some aspect of management. Yes, it takes nerve to do this, but it might land you a job if you are self-confident and have prepared intelligent questions. If no job is forthcoming in spite of all your efforts, pick your ideal company and ask if you can apprentice yourself to someone you can learn from. Several of my students and even older women have ended up with good jobs after they had a chance to prove themselves.

Step three: Get set for the interview

The point of the résumé was to get you an interview. The point of the interview is to get you the job. The interview is your first foot in the door. Doing well in an interview is an important skill. You may want to go through a few even though the job does not interest you, just to practice. If at all possible, deliver your résumé in person and set up an appointment for an interview while you're there. It will give you a chance to talk to a few people so that next time you come, you will be already a familiar face and the surroundings won't be so strange to you. Find out who will be interviewing you.

If your appointment is with Personnel, this is only a screening process, probably the first of at least two steps. They usually do not do the hiring. On the other hand,

*I gratefully acknowledge the help of Barbara Levin, graduate assistant.

JOB-HUNT CHECKLIST*

I have identified my assets in terms of skills, abilities and personal resources.

I have identified my interests.

I have thought about the type of life-style I want.

I have at least mentally summarized my personal experience in terms of work, education, and vocational activities.

I have identified personality characteristics that should be considered in my job decision.

I know my feelings about the environmental setting I would like.

I have researched (at least generally) various career possibilities.

I have narrowed my career choices down to a manageable number worth further exploration.

I have identified employers or people relevant to my career possibilities.

I have talked to one or more persons in the field(s) I am considering.

I have narrowed down the actual job titles I am considering.

I know the various entry-level positions in the field(s) being considered.

I have identified any additional educational or experiential background I should get or can get to better prepare me for or help test my choices (relevant courses and workshops, graduate school, summers of volunteer or paid experience, relevant extra curricular activities).

I have taken the necessary steps to obtain this additional background if necessary.

I have obtained employer and professor recommendations.

I have prepared a good résumé and know how to write cover letters and thank you notes.

I am familiar with the interview process.

I have identified sources of assistance and other pertinent resources.

I have identified some employers I want to contact for interviews.

I have, at least minimally, researched these employers.

I have an understanding of the various approaches used to obtain interviews.

I know what I want and how I'm going to go after it.

I have reasonable alternatives in mind if unable to initially get my first choice.

I have talked over my best alternatives with significant others (mate, friends, parents, and others, etc.)

---

*Gerald D. Brody, Director, Career Development and Placement Center, Kent State University, Kent, Ohio, Reproduced with permission.

they have a lot of power, because if they do not see you as a potential candidate, you will not be considered for the job. If you are interviewed by a specific person for a particular job, it is probably that person who will make the final selection, and it will be up to you to convince that person of your competence. Organizations use several types of interviews, and you should be prepared for them:

The Panel Interview You will be facing several people at once, all asking you questions. Treat each person as an individual. Remember that they do not expect you to leap up with answers to all questions. Take your time to reflect and you will be seen as a thoughtful person.

The Successive Interview You will be seen by several people one *after* the other. It may be an exhausting process; similar questions tend to be asked over and over again. You must answer each one as if it were fresh. Some of my students have spent a full day in successive interviews without even a break for lunch. However, if this happens to you you will do better if you ask to have one of the interviews over lunch. You will be seen as unafraid to ask for legitimate consideration.

The Group Interview Several applicants are seen together by one or more interviewers. The trick is to stand out from the crowd, yet not be seen as overbearing. Have a couple of thoughtful questions ready. The way you balance taking up too much air time and taking up too little will be evaluated.

The Assessment Center This is a simulation exercise that can last anywhere from half a day to three days. A group of candidates is given a task that includes decision making, priority setting, planning, and organizing all under the stress of a time bind. The task is usually in the form of an in-basket full of papers to sort out and act upon. The evaluation is based on your ability to act quickly on some items and to take the time to inform yourself better on others.

Before your interview, find out as much as possible about the company or organization. Get a copy of their annual report and any brochures they offer. Look them up in *Standard & Poor* or any other appropriate publication. Then see if you can answer the following questions:

1. Does this company have a good reputation? a stable financial situation?

2. Does it have a growing market? Is it therefore expanding its personnel?

3. Is it diversified or does it depend on one product or one type of service?

4. Do you believe in the products or services it offers?

5. Are there plans for new products or services?

6. Is promotion by seniority, by merit, from within? Are women being promoted? Are women's salaries equal to their male peers'?

7. Is there an Affirmative Action Officer? an Equal Employment Opportunity Program?

8. Are women included in the training programs?

When appropriate, use this information during the interview. You will be seen as a professional who has made the effort to find out about the company. Also, this process will help you understand the organization and ask the right questions in terms of your career goals. It will familiarize you with the language of the particular business, for each profession has its own jargon.

After you have done your pre-interview homework, plan what you're going to wear.

Because 80 percent of all communication is nonverbal, the way you appear conveys important information to your interviewer. Your physical appearance gives the message as to how you want to be seen. Unless you're applying for a job in a very innovative and creative organization, wear something that will be reassuring to your future employer. If in doubt, be conservative. Wear nothing too short, nothing too low, nothing too tight, nothing too bright. You want to give the appearance of being a serious worker, a serious professional. Be as attractive as you can, but do not be flashy. This is not the time to be "true to yourself" and wear your favorite hippy outfit. First impressions are critical. I like to wear suits for my important first meetings. Research shows that men are most likely to hire women who wear jackets.

Be prepared to deal with stereotypes in your interviews. Despite the fact that many organizations are actively recruiting women and have strong affirmative action plans, there is a good chance that you will face some preconceived notions about what women can or cannot and should or should not do. You may be placed into specific slots of what "all women" are like. Knowing the possible perceptions of others gives you a way of dealing with them: Confront them directly, emphasizing the positive aspects and minimizing the negative ones. The following chart gives you some idea of what you may be facing.

To overcome the stereotypes, you need to develop a strategy to convince your interviewer of your competence, to give examples of your *initiative*, your *responsibility*, and your *enthusiasm*. Two indices of potential managerial ability are *self-confidence* and *energy*. Exhibit appropriate levels of both. When you are interviewed by the person who can hire you, it is important to be able to tell that person, as specifically as possible, how you can be helpful to him or her and to the organization. Knowing what to expect helps.

There are two basic types of interviews:

Directive The interviewer leads the interview, asking most of the questions. This is the easiest interview if you have a fairly typical background. If you do not, it may be hard to place a word in edgewise and to demonstrate your special competencies.

Nondirective The interviewer lets you take the lead and interjects very little. It is the easiest if you are comfortable about marketing yourself and your abilities. It is difficult if you get tongue-tied. Have a few notes about yourself ready to fall back on.

Typical interview questions A usual starter for interviews is "Tell me about yourself" or "Why are you interested in this position (or field)?" You need to have thought through and practiced the answers to these and a variety of other questions. Here is a list of ten frequently asked questions in interviews. Be prepared to answer them.

1. What are your major strengths?
   Do not be modest. Talk about your strengths with assuredness but without cockiness. Perhaps a quiet confidence would be the right expression.

Stereotypes

Entry: The Recent Graduate	Re-Entry or Late Entry: The Homemaker
Enthusiasm of youth	Wisdom of maturity
No life experience but knows latest development in the field	Rich life experience but is not aware of latest development in the field
Has credentials; can brandish a diploma, a list of courses taken, and summer job experiences	Has competencies but does not know their value; does not know how to brandish a list of volunteer experiences
Seen as flexible, able to learn	Seen as rigid, set in her ways
Seen as a risk: may marry, have children, get transferred with husband	Seen as stable and reliable
May be ambitious; will demand equal pay	Will be grateful; will accept lower pay
Because of affirmative action, may have good opportunities	Because of age, may be discriminated against
May be discounted because of youth	May be discounted because of age
May job-hop; sees herself as having options	Will not move; does not see other options
High self-esteem; optimistic; expects to be promoted fast	Low self-esteem; pessimistic; does not expect promotions
Wants to travel	Wants to stay put
Willing to work late; has higher energy level	Willing but unable to work late; gets tired quickly
Has no family commitments	Family commitments may interfere with work schedule (unless family is grown)
Will be seen as daughter or sex object	Will be seen as mother
Vertical orientation: socializes with bosses	Horizontal orientation: socializes with co-workers

2. What are your major weaknesses?
   Remember that weaknesses can be seen as strengths, such as being a perfectionist or always needing to understand the total picture.

3. How is your previous experience (or education) applicable to the job you are seeking?
   If you have done your homework, you will have thought about the applicability of your knowledge and skills to what the company needs.

4. Why did you leave your former job?
   Never speak badly about your former teachers, colleagues, bosses, or company, even if you left on bad terms. Talk instead about the limited fit between your career aspirations and the company. Loyalty is an important attribute and you may be seen as disloyal.

5. What are your career goals?
   If you are not sure of your career aspirations, just talk about wanting to do the best possible job.

6. What are your interests outside of work?
   It is important to list at least one outside interest, possibly in community affairs. The interviewer is trying to put together as complete a profile as possible in a short time. Your hobbies will give a clue as to the kind of person you are.

7. Are you applying to other companies?
   If you are applying to other companies, say so. The interviewer will see you as an honest person—and will see the market for your services as more competitive.

8. What salary do you expect?
   The salary question is tricky. Do not answer unless you know what people at your level are getting and what your minimum requirements are so that you are prepared to either accept, decline, or negotiate an offer. Ideally, defer the salary issue until you have a firm offer.

9. Why should we hire you above all of the other candidates we are interviewing?
   The answer to this is that you do not know the other candidates and cannot possibly tell in what way you might be superior or even different. However, you would like the interviewer to be aware of your skills, competencies, and desire to work for the company and then make that judgment.

10. How fast can you type (asked only of women). If you do not wish to use this skill in your job don't say how fast—or slow—you are. Rather, say that you don't see its relevance to the type of job you are applying for.

Some interviews are designed specifically to put you in a particularly stressful situation to see how you handle it. Sometimes this takes the form of questions that are difficult to deal with and may throw an unprepared candidate off balance. Examples are: Do you feel you have leadership potential? Why? What have been your worst work-related problems? What is your goal in life? Where do you want to end up in our company? How do you relate salary to success? In other words, really tough questions! If you don't have a good answer ready, say it's an interesting or important question and you would prefer taking time to think about it, rather than give a glib answer.

Other stress interviews could take the form of two or more people shooting questions at you, having you sit with the sun in your eyes, or making you wait a long time. These test your ability to assert yourself. Request people to take turns with their questions, move your seat out of the sun.

Certain interview questions are today illegal. You should know what they are. The decision as to whether you answer illegal questions is yours. The law states that you do *not* have to answer, but you should be prepared to respond appropriately if you decide not to answer questions of an illegal or sexist nature. To tell interviewers that they have no right to ask a question may make the interviewers defensive. You may educate them for the next applicant, but you may also be out of a job. A good answer to an illegal question is to ask in what way it pertains to the job. This can be done aggressively or very nicely. Sometimes, if the interviewer is really obnoxious, this may actually represent the company's attitude; you may not want to work there. In this case, a note to the head of the company explaining what happened might be very helpful for future applicants, and you have nothing to lose.

Here are some illegal questions. An interviewer may not ask you:

1. Whether you're married, single, divorced, or living with anyone. This can only be asked for insurance purposes after you have been hired.

2. If you have children, whether you plan to have children, how old they are, or who cares for them. After hiring, the employer may request this information for insurance or other benefit purposes.

3. How old you are, how tall you are, how much you weigh. Nor can any comments be made about your sex or age, unless it is "BFOQ" (bona fide occupational qualification, such as specifying "young women" to model junior fashions).

4. If you have ever been arrested or served a jail term, unless security clearance is needed. Nor can you be asked what type of military discharge you have had or what branch you served in.

5. Whether you live in an apartment or a house, whether you own or rent.

Some of my students have come up with really good answers to such questions. For instance, to "What is your religion?" one student answered, "Oh, does your company represent a particular sect?" To, "Are you married?" the answer was, "Does your company have a preference to hire either married or single people?" To questions dealing with children, you might answer something like, "Is it because you have day care facilities that you're interested?" And finally, to a question that several of my students have heard, "Which method of contraception do you use?" the answer we have strategized is, "Which one do most people in your company use?" In other words, a strategy that is effective but is not seen as overly aggressive is to answer these questions with other questions. Humor may be very useful in an unpleasant situation. On one hand, it is perfectly legitimate and appropriate to just stand up and say that you do not wish to work for a company that asks these kinds of questions and walk out. But then do write the letter to the company president. On the other hand, it is always possible that the interviewer may ask illegal questions innocently, and nothing more is meant by "Are you married?" than to break the ice. You may decide to be reassuring and say "I assume that you are concerned about my long-term commitment. I can assure you that if I get the job I will be as dedicated to my career and to the company as I would if I were single." Or you can simply say, "Yes, I am married."

For more complete information on the new laws affecting women and the workplace, I suggest studying one or more of the following works:

Affirmative Action Glossary, American Association of University Women, 2401 Virginia Ave., N.W., Washington, D.C. 20037.

A Working Woman's Guide to Her Job Rights, U.S. Department of Labor, Washington, D.C., 1975.

"Know Your Rights Under the Fair Credit Reporting Act—A Checklist for Consumers," *Consumer Bulletin No. 7,* Federal Trade Commission, Washington, D.C.

Denise Brinder Leary, *Federal Laws and Regulations Prohibiting Sex Discrimination,* Washington, D.C. WEAL.

Karen Lindsey, "Sexual Harassment on the Job and How to Stop It," *Ms.,* November 1977.

"Surveys Show Widespread Sexual Harassment of Working Women," *Fair Employment Report,* June 13, 1977.

Women and Corporations: Issues and Actions, Interfaith Center on Corporate Responsibility, 475 Riverside Dr., New York, 10027.

Interview the interviewer We have been dealing with the questions that you might be asked, but an interview is a two-way street. You too are interviewing the other person about the job. It is important that you ask some questions because they will aid your own assessment of the job and because they show your serious interest. If you want the job, say so. Companies are more interested in hiring people who are enthusiastic than people who seem indifferent. Here are some questions you should ask:

1. What are your actual duties and responsibilities? Is there an order of priority?

2. To whom do you report? Who reports to you? Who are some of your colleagues?

3. Is there a training period? What does it consist of? Is there a probationary period? How long is it and on what are you being evaluated?

4. Will you be working evenings, weekends?

5. What are the chances of promotion?

6. How flexible is your supervisor in terms of hours?

7. Are you expected to travel?

8. What is the situation with parking. Are there any car pools or is public transportation available?

9. Is your job exempt or nonexempt?* (To be asked for entry-level jobs only.) What are the benefits provided?

10. What will your salary be? What are the chances for increases? How frequently? Note: Do not ask about salary, benefits, vacations too early. You may want to wait until you have a job offer.

In addition, ask questions about the current organizational objectives, the future goals, the hopes and concerns the interviewer has for the company, and in what way you might begin to fulfill these needs. For example, it is better to say ''I can write ads that will double your response'' than to say ''I'm a good advertising copywriter.'' A candidate must show both her ambition and her concern for the company's welfare. A real fit is established when you meet your needs through your job and the organization meets its needs through you.

It is up to the interviewer to close the interview. However, if you feel you have said all that is necessary and received all the information you need, you may say something like: ''Well, how are we doing? I am satisfied. Are you?'' It is legitimate to then ask what the next steps are in the hiring procedure and what your chances are.

It is important for you to make a list of these questions and others you may have and to bring the list with you to the interview. Otherwise, in such a stressful situation, you may forget to ask some critical questions.

Before leaving, ask the interviewer when you may call to ask about the status of your application. I prefer having that control. There is nothing more nerve-racking than to wait for a telephone call without a definite time limit. If your interviewer insists on calling you say, ''OK, but if I haven't heard in a week (or month or whatever time you have agreed upon) I would like to call you.''

After *every* interview, even after a telephone contact, write a thank-you note. Remember what your interviewer seemed to respond to positively and negatively and what you found out about the company and the job, especially about the qualifications that seem to be needed. Then write the letter as soon as possible, including in it your special competencies that fit the job as you now see it, adding any additional appropriate skills you feel your interviewer should be aware of. If you met anyone else, mention them by name in a complimentary manner; comment on how much you enjoyed your discussion with your interviewer; most important, reiterate your interest and enthusiasm for working for that company. Be sure you know the correct spelling of your interviewer's name, and of his or her title.

*Exempt means that you are not paid overtime; nonexempt employees are paid an hourly wage and *do* earn overtime pay.

If you do not get the job, make an effort to find out why. Even though this may not be particularly pleasant, finding out may help you with the next interview. Some interviewers will help you, others won't, but it's worth the try. However, do not assume that you weren't hired because you did not fit the job or impress the interviewer. Too frequently, jobs are advertised only to comply with affirmative action regulations, whereas the job has already been earmarked for an inside person. So instead of being discouraged, think of it as another opportunity to practice your interviewing skills. But there may be another reason why you did not get the job!

THE CLONAL EFFECT

You have identified your competencies; you have written the perfect résumé; you were superb in your interview. And you were not hired. What's wrong with you? Nothing! It's the clonal effect! What is the clonal effect? It is the tendency of individuals, groups, and organizations to replicate themselves or others that are close to them, whenever they have an opportunity to do so. The dictionary defines a clone as a "group of organisms descended asexually from a single common ancestor."

Every time someone is to be hired or promoted and there is a pool of available candidates, there are two criteria that enter into play. One is competence to do the required job; the other is the fit between the individual to be hired and the rest of the staff and organization. This is where the clonal effect takes place. The "fit" deals with the comfort level the employer or employers feel with the person being hired or promoted. "Fit" is in the eyes of the beholder.

Recent research findings point to the fact that, when people are discriminated against because of gender during the hiring process, it is often in terms of imagined fit. Men are hired more often to manage male subordinates, and women are hired for female subordinates.[1] In other words, fit deals with the unconsciously perceived comfort levels of the subordinates, who are able to relate better to a manager of their own sex. I am not saying that this perception is erroneous. It is important for us to note, however, that females will be hired less often as managers in male-dominated organizations and that this will be an unconscious discrimination.

An employer hires someone with whom he or she has a fair chance of getting along, of communicating well, or sharing basic values around such matters as work ethics, standards of quality, imagination, precision, punctuality, dress codes, humour, politics, and even leisure activities. The list is endless and so are the possible prejudices. Whom do we trust? Those whom we can understand, those who are most predictable to us. Who are they? Those who are most like ourselves.

Research conducted with 207 male and 72 female managers "confirms that if the managers perceive people as like themselves, they tended to like them (or if they liked them, they considered them to be like themselves)."[2]

People joke about bankers all looking alike. Well, if you don't look like one, you won't get hired for that job. People talk about the Madison Avenue type, the cloak and suiter, the salesman, the social worker, and everyone knows the images conjured by these stereotypes. Stereotypes perpetuate the species, for each will keep cloning itself until stopped by their own awareness of doing just that. In the midst of hiring a new assistant professor at my university, I suddenly realized that our most dynamic professor felt that the candidates were too quiet and wondered if they were depressed. Our most quietly thoughtful colleague worried about their hyperactivity. The one I liked best just happened to be short and freckled like me.

The clonal effect does not stop at reproducing ourselves, we also tend to clone those who were or are close to us. In choosing a mate, people unconsciously select those who have a physical and/or psychological resemblance to a parent or sibling. Frequently, the youngest boy of a family will tend to select as a wife a woman who has a younger brother, and she in turn will look for someone who has an older sister. The family positions tend to be replicated in marital relationships because of the familiarity and comfort we recreate for ourselves. People clone the sibling into the spouse.

Just as individuals and families tend to replicate themselves, so do groups and organizations. The tendency is to replace lost members with people who have similar characteristics or to add people who would not change too much the dynamics of the usual communication patterns. Even in the effort to add to a work unit people who have complementary skills or knowledge, a certain comfort level of predicted interactions has to be maintained.

We often talk of race, sex, and ethnic origins as bases for discrimination. We do not as frequently discuss socioeconomic class. Yet class is as much a factor in selecting mates, friends, and colleagues as the more publicly evident characteristics. Because class is more difficult to define, the cues are subtler and discrimination is based on the identification of often minimal clues: a walk, a choice of words, a piece of clothing, a look, a joke, a mannerism; the way someone enters a restaurant, orders, eats, pays the check, relates with the waiter. Most of us recognize class. The prejudice, of course, is not in the acknowledgment of the difference, but in the *preference* of one over the other, and the discrimination is in *acting upon this preference*.

Again, the tendency is simply to be more comfortable with the person who comes from a similar background, to be more trusting of those who are from the same class. In hiring and promotion, classes tend to clone themselves, too.

In order to become conscious of the clonal effect, in order to have more choices, you need to point it out, talk about it, even joke about it. If we wish to stop the unconscious tendency to reproduce ourselves, then we must actively look for the discomfort of diversity, the challenge of change, the potential for disagreement. Then and only then, will the women, the blacks, the people of different religions and various ethnic backgrounds, the handicapped, the too young, the too old, the too skinny or too fat, the oddly dressed, and those with strange accents have a chance to join in so that all of us can contribute our differences and be enriched by them.

How can you deal with the clonal effect if it is mainly unconscious? Address it directly. For instance, if you ask your prospective employer whether most of the employees have the same background, or come from the same area, or seem to all share similar interests, or are all white Protestant males, he or she will have to think a moment before answering. You can then suggest that "diversity is creative and healthy for a company; however, it is difficult for many people to go beyond the difference and see the potential resources." If you are not Protestant, anglo-saxon, white, or male, you have something special to offer—another point of view, another reference system. If you value your "otherness" and offer it as a necessary ingredient to the "sameness" usual in most companies, you will have a good chance to be seen as an important addition.

MAKING A CHOICE

Your résumé got you interviews. The interviews got you job offers—yes, several. How do you know which job to take? Go back to the list of questions you should ask the interviewer (pages 43 to 44). Look at the questions regarding salary increases and promotions, but now be very specific: How frequently are promotions made, and based on what criteria? Have other women been promoted? What is the upward path for most people in this company? What about benefits and vacations? What are the sick-leave and maternity policies? Does the company have a bonus system? a profit-sharing plan?

After you have all your questions answered, do the most important thing of all: *Take a good look around*. Do the people seem happy and productive—not too harried but not bored either? Who will your co-workers be? Can you meet some of them? Trust your feelings. You will *feel* the atmosphere and with your knowing, will react with either "I would be happy working here" or "It doesn't feel right." This is important data; it means you're picking up something you should pay attention to.

Before you start

You are now in that time between having accepted a job and showing up for your first day at work. Examine carefully your expectations. Young women just out of school sometimes expect to jump right into responsible positions. Older women returning to work may come in very tentatively. Each should learn from the other. The young assertive woman with high expectations will have to lie low at the beginning, learn the organizational norms, be patient, and pay her dues in terms of time before expecting recognition. The older, diffident woman will need to believe the company is lucky to have her and not remain so invisible that she is not recognized for her work.

Whichever is your way, stay in touch with yourself, your feelings, reactions, needs, and expectations as you work on fitting you with "them." You will need to learn not only about you but also about "them." For only when you understand your own needs as well as others' expectations within the context of organizational requirements will you know whether there is a good fit—or at least the possibility of one.

NOTES

1. Gerald L. Rose and P. Andiappan, "Sex Effects on Managerial Hiring Decisions," *Academy of Management Journal* 2(1978):104–112.
2. John Handyside, "Perception," *Psychology Today,* November 1979, pp. 24–26.

SELECTED READINGS

Berscheid, Ellen, and Walster, Elaine. *Interpersonal Attraction.* Reading, MA: Addison-Wesley, 1969.

Bolles, Richard Nelson. *What Color is Your Parachute: A Practical Manual for Job Hunters and Career Changers* (rev. ed.). Berkeley, CA: Ten Speed Press, 1977.

Cohen, A. R., and Gadon, H. *Alternative Work Schedules: Integrating Individual and Organizational Needs.* Reading, MA: Addison-Wesley, 1978.

Epstein, Cynthia Fuchs. *Women's Place: Options and Limits in Professional Careers.* Berkeley, CA: University of California Press, 1970.

Frank, Harold U. *Women in the Organization.* Philadelphia: University of Pennsylvania Press, 1977.

Hall, D. T. *Careers in Organizations.* Santa Monica, CA: Goodyear Publishing, 1976.

Hansen, L. Sunny, and Rapoza, Rita S. *Career Development and Counseling of Women.* Springfield, IL: Charles C. Thomas Publishers, 1978.

Herman, Deborah D. "More Career Opportunities for Women: Whose Responsibility?" *Personnel Journal,* June 1974.

Hopke, William E. (Ed.). *The Encyclopedia of Career and Vocational Guidance.* New York: Doubleday, 1974.

Irish, Richard K. *Go Hire Yourself an Employer.* New York: Anchor Press, 1974.

Josefowitz, Natasha. "Ahmad vs. the Group: On Managing Differences." *Social Change* 9–1(1979).

Josefowitz, Natasha, and Gadon, Herman. "Evaluation Report of the Women's Career Project." Boston: Northeastern University, 1978.

Lathrop, Richard. *Who's Hiring Whom.* Berkeley, CA: Ten Speed Press, 1974.

Levitin, Theresa E.; Quinn, Robert P.; Staines, Graham L. "A Woman Is 50% of a Man." *Psychology Today,* March 1973.

Lovejoy, Clarence E. *Lovejoy's Career and Vocational School Guide.* New York: Simon & Schuster, 1974.

Occupational Handbook. Washington, D.C.: U.S. Bureau of Labor Statistics, 1974–1975.

Reis, Susan L. *A Woman's Job Search: Five Strategies for Success.* New York: American Association of University Women, 1978.

Rosen, B., and Jerdee, T. H. "Influence of Sex Role Stereotypes on Personnel Decisions." *Journal of Applied Psychology* 59-1(1974):9–14.

Schwartz, Felice N.; Schifter, Margaret H.; Gilloti, Susan S. *How to Go to Work When Your Husband Is Against It, Your Children Aren't Old Enough, and There's Nothing You Can Do Anyhow.* New York: Simon & Schuster, 1972.

Senger, John. "Managers' Perceptions of Subordinates' Competence as a Function of Personal Value Orientations." *Academy of Management Journal,* December 1971, pp. 415–423.

Chapter 3

MY BOSS WINKED AT ME

My boss winked at me today.
First I was pleased—
it's a friendly thing to do—
and then I wasn't.
Would he have winked at a man?
NO
A wink is something men do
mostly to women,
like little pats
wherever.
Does it mean I'm cute
but not to be taken seriously?
Does it mean he likes me
but won't trust me,
won't give me responsibility,
won't promote me,
won't give me raises?
Hmmm . . .
Tomorrow I'll wink at him.

Moving In

Congratulations! You have written a résumé that led to an interview, the interview was successful, and you have been hired. How could you be best prepared for that first day on the job?

If you're just out of school, it would be desirable to have found an apartment and bought enough furniture to be comfortable before you start work. If you're already settled, take the time to do as much of the shopping for clothes and household items as you can before your first day. You will feel *very* tired the first few weeks and will need to protect yourself from any extra responsibilities. Be sure to delegate some of your former tasks to others (roommates, husband, children) *before* you collapse under the burden of attempting "life as usual" plus a full-time job.

Many women, feeling guilty about going off to work and leaving the house, will retain their household-related responsibilities and shift them to the end of their day, or get up at 5:00 A.M. in order to do the laundry. Unfortunately, single working mothers have little choice in these matters while their children are still small. But those who have an alternative merely demean themselves and their families by turning into round-the-clock work horses and not insisting that all members of the family share equally in the tasks. An exhausted woman will not be a good companion, an attentive mother, or an efficient worker.

Plan your clothes and accessories the night before so that you don't have to start your big day worrying about anything except getting there feeling in charge of yourself. Wear something simple and classic until you find out what others usually wear. Assume you will be nervous and sweaty. If it's winter, the offices may be overheated; be able to take off a sweater or jacket. If it's summer, the air conditioning may be set on too cold and you'll need a thin sweater or jacket to keep from catching cold. On the first day take along some crackers and cheese or an apple in case you get hungry from exhaustion or nervousness. It will give you quick energy. Remember that many bathrooms do not have tampon dispensers. Take a magazine (*Business Week*) or a newspaper (*The Wall Street Journal*) with you, in case you are not given anything to do for a few hours, so you won't just sit with folded arms.

FIRST IMPRESSIONS

So here you are! You have arrived and are working on the first day of your new job. You want to make a good impression—this is critical since it tends to linger even if your behavior subsequently changes. First, try to learn your surroundings in order to get comfortable. There are a couple of things you can do to make it easier for yourself. If your boss doesn't introduce you to your colleagues ask for the introduction. Ask whom you can join for breaks and lunch. Find out who can answer any questions you might have and meet that person as soon as possible. Take some time to sit down with your supervisor and make sure you know what is expected of you that first day.

Ask questions—lots of questions: "The curse of the ignorant is their ignorance."* So become knowledgeable about what is going on around you, for *knowledge is empowering.* Do not share too many personal matters; you do not know how it will be interpreted. Be friendly but not too familiar; it is easier to become more intimate later than to distance yourself once you have been closer. Besides, when there is no other information available, people tend to attribute power to those who exhibit more distant and formal behaviors than to those who are immediately warm, friendly, and informal. The latter may be liked sooner, but initial respect may be more important if your aim is to move up from your entry-level position.

STEPPING FORWARD

Although you may have come in as a learner, it is important to not play that role exclusively or you will be seen as dependent, needing help. It is easier, of course, to continue in the learner role because then you are not responsible for errors; you can rely on the excuse "I didn't know any better as yet." But you *must* give up that attitude and show knowledge and strength; you must become accountable for your actions. Whenever our own MBA students graduate and start teaching at our university, they are seldom given colleague status in our eyes. They remain "our students" and have trouble transcending that role. They must teach elsewhere to be seen as credible by their peers. The same is true of an entry-level job: It is difficult to move on and leave the trainee role behind. Upon entering a new organization, the best way is to keep your distance in order to ascertain the situation. Of course, you can and should ask for help, but do it as an adult who needs the orientation, not as a helpless child who is lost in new surroundings.

One issue for women at the entry level focuses on *membership* and *inclusion,* a need that people have in varying degrees. Some people need to take their time and observe; others need to jump right in. How quickly do you need to feel a part or a member of any group you enter? How comfortable are you to remain on the periphery, and for how long? Do you tend to become overly friendly too quickly in order to become known and to know others as soon as possible? I do. Or do you pull back, staying in your own corner until someone else makes the overture? Either way

*From a lecture by Paul Samuelson, M.I.T., 1948.

may be inappropriate. You need to assess the situation. Whatever you do, be aware of your behavior and of the responses it elicits from others.

These responses are part of your data. Are the people formal or informal? Is there frequent contact between people? Is that contact only horizontal (among peers) or is it also vertical (with bosses and subordinates)? Observe carefully and make mental notes of what you see. Although it is true that, as you act, people will react, there is another facet of human behavior not within your control: People will sometimes react in certain ways not because of anything you *do* but because you *remind* them of someone they know. Whether you're tall or short, spontaneous or reflective, you will be attributed qualities and faults that do not belong to you. Age, for example, elicits specific stereotyping.

Age stereotyping

The tendency to react, especially at the beginning, to age stereotypes should be taken into consideration. If you're very young, some men will tend to make a daughter out of you and act in paternalistic ways. If you refuse the role and act independently, you will frustrate the father in them and will be seen as ungrateful. (After all, all he wanted was to protect you!) The hidden message is that you *need* protection, which makes *you* weak and *the men* strong.

If you're somewhat older, you might be seen as a sex object. Some men will make seductive remarks and flirt, either with or without the intent to sleep with you. If you refuse the role and act professionally, you will be seen as a castrating bitch rejecting the masculine ego. (After all, you should feel complimented that he finds you attractive!) The men may not understand the demeaning aspect of being seen as *something* to play or sleep with, as opposed to *someone* to work with.

If you're a mature woman, you might be taken for a mother. Some men will feel so comfortable with you that they tell you their problems, seeking nurturance and acting as dependent children. If you refuse this role and act as a colleague you will push away the child in them and be seen as cold and unsympathetic. (After all, he was just extending his friendship to you!) The men don't behave this way with other male colleagues. They are unaware that this treatment does not place you on an equal footing but gives you a role that can be discounted.

With time, age seems to matter less. But of course, some men will always react to women as their fathers, lovers, or sons, no matter what the age. Many, however, will be able to go beyond the stereotypes and respond to a woman's professionalism and competencies and to her offer of colleagueship.

Sexual overtures

The father-daughter or son-mother types of interactions are not conducive to *collegial* working relationships, yet they are still easier to handle than the lover-mistress relationship. Women as sex objects in our society is so pervasive, so familiar a topic, so frequently joked about, that to take a stand against it is taking a stand against a deeply ingrained American male cultural pattern. It takes the form of inuendos, of remarks about female physical attributes, of stories of how they made out last night, of touching in possessive or sexual ways, and of endless anecdotes. If you don't laugh *with* them, you have no sense of humor; if you *do,* you are disloyal to your sex. What I do is shake my head and say "Oh, what an awful joke," or "You are terrible," or "I refuse to laugh." But I say this good humoredly, not punitively. It is not just *what* we say, but how we say it. Style is the critical component here. You can express your anger if you are a member of the group, but if you are still working on being included, anger will distance you and make it that much harder to become one of them.

Again, we're in a double bind. If we don't respond as a daughter, we are uppity; if we don't respond as a mother, we are cold; but if we don't respond as a mistress, then we are bitchy!

Although the married woman may also be prey to sexual harrassment, it is often the single woman who is seen as available (and therefore willing with just a little prodding) to have an adventure. These expectations of sexual favors can range from a "quicky" in an empty office to weekends together. If she complies, out of fear of losing her job or because of her own needs, *she* is devalued as a professional by her peers and bosses, whereas *he* is not. "Sleeping around" may help at most one step up the career ladder, but it never helps in the long run because in most people's minds, a female sexual partner is not synonymous with managerial abilities. This comes from old stereotypes of "what women are good for." The part that renders women vulnerable is that they may indeed be attracted to a particular male at work. If she acts on her desire she is seen as unbridled; if he does he is seen as normal.

It is this perception of what is normal for a man that makes the situation so difficult. It is "normal" for a man to make advances, to pursue, to crack jokes and make sexual innuendoes, all toward an uninterested and unwilling female. It is not "normal" for her to be upset, angry, outraged, turn him down, or complain to the boss or the affirmative action officer. There is an assumption in many men's heads that (1) women want it; (2) women can't resist when a man insists because she must be flattered by his interest; (3) she was flirtatious and led him on (after all, she did smile at him once and wore that sweater).

It is surprisingly difficult to make a man believe that, although he is not appealing as a potential sexual partner, there is nothing wrong with him as a person or even as a man; the two are equated in many men's identity. In order to deal with sexual harassment, it is important for women to understand that many men come from this position. If we see it as offensive and they see it as a compliment that fulfills our secret desires, then something is wrong with our communications. So the first thing is to let the man know that you consider his behavior an insult—not a compliment—and that you live by a hard and fast rule: You *never* have affairs with your colleagues and you *never* deviate from that rule.

If the harassment continues, ask the man what you should do to make him stop. Reiterate that you are *not* flattered by his attentions, that instead you feel demeaned by them. State that you are not rejecting the person, but the behavior. If nothing works, then you have no alternative but to threaten exposure. If your boss is the culprit, this may cost you your job. You can, however, promise to say no more about it if he is willing to stop bothering you. Fighting off a strong man alone is difficult because some men are excited by a struggling woman and feel challenged to overpower her. If possible, do struggle—scream and run if you can. If the harassment still does not stop, tell him that you are ready to talk to his boss, or to his wife. Secretly wiring yourself with a hidden tape recorder and threatening to expose him afterwards may even be necessary. If he has no boss that you can go to, then you can threaten with newspaper exposure or legal action.

The real problem is when you are threatened with being fired or not promoted if you don't accede. All you can do is threaten a law suit. However, if you don't want to carry through on your threat and are seriously suffering, yet for economic reasons cannot leave, then enlist the help of all possible colleagues in the organization. Even if they are all males, you may find at least a few sympathetic ears, if you talk with them individually. Threaten to go to the human rights commission, to the CEO or Affirmative Action officer, or to the Alliance Against Sexual Coercion (P.O. Box 1, Cambridge, MA 02139), or the Working Women United Institute (593 Park Avenue, New York, NY 10021).

The men who harass women are too often protected by the women's unwillingness to talk. Women should tell each other and form a strong supportive alliance against such men, in order to strategize or go as a group to complain about them.

Whichever role you are cast into, if it is seen as a "special relationship" it may elicit envy or suspicion on the part of your co-workers. The only thing you can do is be *very clear* about your own signals. If you do not act cute, seductive, or protective, but professional and collegial, the suspicion will lessen. However, if you are successful, the envy may remain. This does not mean that you cannot be charming, or fix a cup of coffee, or listen to a problem. Women do this well; it is a part of our strength. But there is a big difference between doing all of these things sometimes and being cast in the role of *only* able to perform the "feminine" or subordinate tasks.

Emotionality

Are women indeed more emotional than men? Men think so. It depends, of course, by what is meant by emotionality. Anger is an emotion that men express more directly than women. Crying is an expression of emotion that women generally use more than men. So what is *really* meant by the commonly held belief that women are more emotional than men is that women are more prone to tears. Recent research on gender differences in emotionality shows that in entry-level jobs, many women do tend to express their feelings of anger and hurt by crying, while most men express these feelings by slamming doors or raising their voices.[1] At the managerial levels, however, this difference disappears: Women have learned to display direct anger instead of crying. These findings seem to point to the fact that women with managerial aspirations will do well to control their tears early on and translate their upset into direct displays. Hopefully, with time, both sexes will be able to cry when that is appropriate.

However, other research seems to indicate that women respond to a wider range of stimuli than men, which seem natural if our senses are indeed more finely attuned. It is wonderful that we can feel so acutely and express so well. How much richer our lives are for it. But we are an enigma to many men who do not perceive the world in the same way. Our reactions to events they don't notice seem to them to be over-reactions, and we are therefore labeled *over-emotional*. The advice has been for us to feel less, show less, be less sensitive in order to fit into the man's world as perceived by men. Instead of narrowing our range, we should consider sensitizing men to feel more and show more in order to broaden *their* range.

This won't be accomplished overnight. It takes patience and dedication, and it must *not* be done angrily. In the meantime, we can function somewhere between being our *full* selves and being what others *expect* us to be. We must be what will work for us best in whatever situation we are in. Even if it looks like we are giving up our integrity, it is only a survival strategy. If we remain in touch with ourselves and on the alert for every opportunity to educate those around us, we are on the right path.

Women can be so much. There is such a wide range of behaviors open to us. As long as we are not cast into narrow definitions of womanhood, we can indeed be whole.

NORMS

Before you can determine which behaviors are correct in your place of work, you need to know the norms of the organization. Until you know what is expected you cannot make the choice of conforming or not conforming. Not knowing is being in a powerless position.

Norms are the unwritten rules stating what people should and should not do. Although they deal with behavior, norms also establish what people should or should not *be,* which deals with attitudes. In other words, a norm is an expectation of what a good member of a group or organization should be like. Norms lend stability and predictability to organizational life, and although they are neither good or bad in themselves, they may be functional or dysfunctional for a particular purpose. They impact on the effectiveness of individuals and groups.

What are some of the more typical norms? An easy one to detect involves smoking. There may be rules about smoking, or only norms—it may even be a norm to break the rules! Other norms include the kinds of clothes people wear, whether beards and long hair are okay, whether people can interrupt each other spontaneously, and whether people must work a full day or can relax and take breaks. Is profanity used? How about joking? Is it okay to change seats at regular meetings, or does everyone stake out a particular territory? Is the norm to look busy, or to look unhurried? Some norms are formed through conscious decisions, such as when everyone always comes on time to meetings; others are totally unconscious; still others are practiced with varying degress of awareness. There may even be a norm not to discuss norms.

The norms that are created and enforced by members of an organization are not always related to all people in all places at all times. They vary in intensity and in the range of acceptable behavior. For instance, it may be okay to take off your shoes when meeting with the people from your own department, but not okay when meeting with people from other departments. Talking about the boss may be permitted in certain circumstances but frowned upon in others. You may be able to talk about the drinking you did last night, but not about the feelings you have for a partner. You can discuss the stock market but not your personal finances.

A leader usually follows the major norms and sets the example for those below. The leader is not punished if he or she deviates from the norms, whereas other members usually are. New members are taught the ropes and are initiated into the rituals and ceremonies of the organization. If you look at the line in the following diagram, the left-hand side shows the rigid norms that do not bear deviation. But too much security can result in monotony, boredom, oppression. The other end of the line indicates a lack of patterns for behavior. The unpredictability and challenge can breed anxiety.[2]

Well-defined Norms   Ill-defined Norms

Security═══════════════════════════════════════**Challenge**

(monotony, oppression)   (unpredictability, anxiety)

An example of a rigid norm would be that no one could come to a meeting late and would be sanctioned for it without exception. On the opposite pole is the case in which no one knows for sure when to arrive because no one else ever comes on time.

How do you figure out the norms of your department or the organization? They may not always be identical. It may be all right to extend lunch breaks in your unit but frowned upon in others. You may be very informal with your boss but need to be formal with other managers. The only way to know is to be attentive to everything that goes on around you. Be on your toes and make continuous mental notes about what you see. Very few behaviors (if any) are meaningless or out of context. Observe patterns of recurrent behaviors. What seems to get rewarded? What elicits disapproval? Do as others do until you are an accepted member. Then you can attempt diverging from the norms. Here is a list of behaviors you can look for in order to ascertain the norms of your company.

- People all dress similarly/individualistically.
- People are treated formally/informally.
- Desks are clean/cluttered.
- Doors to offices are left open/closed.
- People arrive on time/arrive late.
- People leave on time/stay late.
- The telephone is used for personal calls/for business only.
- People always stick together/seek to meet new people.
- People wait to be assigned tasks/are self-starters.

- Promotions are from within/from without.

- People write memos/call/see people in person.

- People oversell themselves/are modest.

- People openly discuss salary and personal matters/never talk about it.

- People question authority/always accept decisions from superiors.

- People go through channels/communicate directly.

- People check it out first/assume approval.

- Profanity and joking are accepted/frowned upon.

- People collaborate/compete.

- People are patient/pushy.

- People ask for help/figure it out on their own.

- People confront conflicts/sweep them under the rug.

- People take on more responsibility than assigned/just do their own work.

- Women are not accepted/are welcomed at men's informal gatherings.

- Women are discounted/are treated as equals.

- Women are not promoted beyond first-line managerial positions/are in top executive positions.

It is critical for women entering an organization to be able to identify and understand the norms. Norms may cause great difficulty for women. For example, men, when they speak to one another, tend to have direct eye contact. For women, the norm may be for them *not* to look the men in the eye when speaking, since it may be seen as flirtatious or too assertive behavior. In other words, if women act like the men, they will anger them; if they don't, they will be discounted.

Add to this list other norms you have observed and then decide which of these are functional or dysfunctional for the company and for you personally. You may be tempted to change some of the more dysfunctional norms. Wait to do so until you are an accepted member of your company or you will risk being seen as a nonconformist, which may or may not be a norm.

NORM CENSUS*

	Functional	Dysfunc- tional	Impact on me
My unit or department:			
My company:			
Conflicting norms:			

*After you've had time to observe the standards of both your department and your company as a whole, identify the norms you've witnessed in the chart above. Determine whether they are functional or not, and write down your feelings about them. Do any norms conflict?

EXPECTATIONS

Women must learn to deal with a set of expectations that is different from that of their male colleagues. Some of these expectations deal with the female stereotypes of usually being reactive and responsive rather than being proactive and initiating. Women are known as conflict avoiders, peacemakers, and approval seekers, so they violate expectations when they confront people or go ahead on their own cognizance without checking for support. A man's masculinity is generally enhanced by success; a woman's femininity is often questioned by it. For a man to become successful, he must be public and assertive. A woman must do the same, yet if she does, she violates the expectations of female passivity and dependence.

The double bind

The double bind that women often find themselves in in organizations is that, to succeed, *they must play the game according to the rules, yet they are not allowed to do so.* Risk taking is riskier for women—but you *must* do it if you wish to be recognized for your competencies. If you don't, you will remain unnoticed.

Impressions From An Office

The family picture is on HIS desk:	**The family picture is on HER desk:**
Ah, a solid, responsible family man.	Umm, her family will come before her career.
HIS desk is cluttered:	**HER desk is cluttered:**
He's obviously a hard worker and a busy man.	She's obviously a disorganized scatterbrain.
HE is talking with his co-workers:	**SHE is talking with her co-workers:**
He must be discussing the latest deal.	She must be gossiping.
HE's not at his desk:	**SHE's not at her desk:**
He must be at a meeting.	She must be in the ladies' room.
HE's not in the office:	**SHE's not in the office:**
He's meeting customers.	She must be out shopping.
HE's having lunch with the boss:	**SHE's having lunch with the boss:**
He's on his way up.	They must be having an affair.
The boss criticized HIM:	**The boss criticized HER:**
He'll improve his performance.	She'll be very upset.
HE got an unfair deal:	**SHE got an unfair deal:**
Did he get angry?	Did she cry?
HE's getting married:	**SHE's getting married:**
He'll get more settled.	She'll get pregnant and leave.
HE's having a baby:	**SHE's having a baby:**
He'll need a raise.	She'll cost the company money in maternity benefits.
HE's going on a business trip:	**SHE's going on a business trip:**
It's good for his career.	What does her husband say?
HE's leaving for a better job:	**SHE's leaving for a better job:**
He knows how to recognize a good opportunity.	Women are undependable.

Women who wish to move up need to become exquisitely perceptive evaluators of how far to push the norms and how fast. We need to tread the fine line between not meeting others' expectations of how women should behave, yet not upsetting people to such a degree that our objective is never reached.

The task becomes a matter not of breaking the stereotypes but of *how* and *when* to do it. For career-oriented women, respecting male expectations of what "female" behavior is like is not only self-defeating in terms of opportunities to advance, but also in terms of opportunities to change norms, modify stereotyping, and diminish discrimination.

Male socialization

Just as we are victims of socialization, so are our male colleagues. The difference lies in the fact that our socialization has proved to be mostly dysfunctional; women are motivated to constantly anticipate and be ready for change. The male's socialization is not as evidently dysfunctioning—in fact, most men have enjoyed it. They have not minded the female subservience—the acceptance of women to do the less important and menial work, to reap fewer rewards, and to be seen as the weaker sex. Men are therefore less motivated to work toward the awareness, understanding, and change that would no doubt affect their life-styles and possibly their job security.

If we realize the importance for many men of keeping things as they are, we will move cautiously, but we *will* move. Do small things at the beginning, such as getting your voice heard during meetings. Start out by asking a clarifying question, then voice agreement, question the validity of an argument, start to take sides, and eventually initiate an idea, disagree, and insist on being heard.

At all times, show interest in the company, its products or services, and its policies. Stay in close contact with what needs to be done. Then as long as it is not below the position you have been hired for, do it—especially if it is a responsibility attached to a level above you.

FORMAL AND INFORMAL ORGANIZATIONAL STRUCTURES

Survival is predicated on the knowledge of the environmental potential to either harm or help. We need both the ability to escape damage and the capacity to extract the resources necessary to not only survive but even thrive. If the environment is so hostile that all we can be concerned with is survival, then we are better off just protecting ourselves by becoming invisible: Do our job quietly and never ask for anything; smile and say "Thank you" a lot; do a lot of favors for a lot of people and ask none in return. Be at all times attentive, apologetic, humble, noiselessly hard-working, and helpful—and always grateful for the permission to hold on to that wonderfully menial underpaid job.

Of course, there is no guarantee that this will keep us there forever, but at least we have a good chance at it. And for many women it may well be that sheer survival *is* at stake. However, if you would like to *master* your environment and are willing and

Symbols of power

Position	Office location	Office space	Desk size; chair	Floor covering	Office decoration
Entry level	different building from headquarters	shared space in large area; no partitions; visible at all times	small metal desk; small metal chair	none	none
First years	distant floor from headquarters	shared space; movable partitions; no doors	large metal desk, arm chair	none	none
Supervisor	floor next to headquarters floor	office shared with 1 or 2 others; if alone, no windows, but with door	small wood desk; swivel chair	none or small area rug	adds a few personal touches
Middle management	same floor as headquarters but distant from it	own office with window	medium wood desk, swivel arm chair	large area rug	adds many personal touches
Upper management	next to headquarters or close to it	corner office, executive toilet	large wooden desk, leather swivel arm chair that leans back	wall-to-wall	secretary helps in decorating office
President or chairperson of the board	whole floor or floor section	office suite, own toilet and shower	antique or custom-made desk and arm chair, sofa	Persian rug or custom-made carpeting	hires decorator to decorate office
Retired	same floor but distant from headquarters	own office with windows	medium wood desk and comfortable arm chair	area rug or wall-to-wall	many personal mementos but no special decoration

Position	Telephone use	Secretarial help	Time autonomy	Male clothing	Female clothing
Entry level	none; sometimes phone on desk	does own typing	must come and leave on time (punch clock)	short sleeves, open shirt, and corduroy pants	slacks, short skirts, casual clothes in bright colors; handbag and shopping bag
First years	phone on desk; dials and takes own messages	use of secretarial pool	must ask boss; very little lattitude	turtleneck and slacks	pantsuits; dresses in polyester knits; large handbag
Supervisors	secretary takes messages but does not screen calls	shares secretary with several others	asks boss for permission	long sleeves; jacket and tie hanging in closet	pantsuits, jacket over hips, no polyester; longer dresses; handbag
Middle management	secretary screens calls	dictates; shares secretary with 1 or 2 others	tells boss	jacket and tie	suits and 2-piece dresses; briefcase and small bag
Upper management	secretary dials and screens calls	own executive secretary; explains what to write	tells secretary	good suits	good clothes in neutral colors; good accessories; attaché case and no handbag
President or chairperson of the board	has private number; secretary makes and takes calls; hard to reach	has executive secretary who also has secretary	comes and goes at will; tells no one	custom-made suits; short sleeves OK	designer clothes and accessories; attaché case or designer handbag
Retired	makes and takes own calls	shares secretary with others	no one cares	appropriate for older man	appropriate for older woman

able to risk some hard knocks, disappointments, and rejections, then you must study it carefully, and study the organization in which you work.

First, obtain an organization chart from your supervisor. All organizations should have them. You will see at a glance who reports to whom, who is in charge of what, and what the formal lines of communication are. The formal structure of the organization as depicted on the chart will tell you the way things are supposed to be, but it will not tell you the way things *really are*. The informal structure of the same organization is a chart you must make up for yourself. You need to find out who is really influential, who has the real power, who is moving up, who controls the resources, who has access to important information, who makes decisions.

How do you find out? Whatever level your job is upon entry, make it one of your top priorities to learn all you can about the organization and about the people in it. Most companies have at least one person who has been with the company for years and who would love to talk about old times. That person, "the company historian," will tell you about how it got started, about the founders or their descendants, about the early products or services. You will begin to gain a sense of what this company is all about. If, for instance, an old New Hampshire farmer was influential in the beginning, his legacy may be that honesty, integrity, hard work, privacy, and a "no frills" system will be what is valued.

Notice how the bosses dress and act and what they do for leisure; observe which of their subordinates dress and act similarly and have the same leisure activities. These people will be moving up. Observe who is friendly with whom, who sees whom after hours or on weekends, who has a company car, who leads the more prestigious projects, who goes to whom for information and advice. Find out if the bosses have relatives among the employees or if any of the boss' spouses are friendly with employees' spouses. Pick up subtle cues of what gets rewarded and what gets disapproved. For example, what happens when someone initiates an idea or disagrees with an upper echelon person? After some months you will begin to see patterns. You may not yet fully understand what goes on behind closed doors, but you will at least know who is there.

What all this data will give you is the informal power structure of the organization, based not on the formal channels only but also on the informal communication network. One of the things you should be looking out for is whether this is a place where you, as a woman, have a chance of being accepted as a professional, or whether sexism is so ingrained and prevalent that only your femaleness will be seen, not your competence.

CHOOSING UPWARD MOBILITY

If you make such a chart for yourself, it will show you who deserves your attention—with whom you should talk, seek out for advice, do favors for, ask out to lunch, and generally be friendly. If you don't like what I am saying because it seems manipulative, you have a choice. You don't have to be political in order to meet your own ends, in this case a professional career; but if you don't do it, be prepared to remain where you are. You may be perfectly satisfied and there is nothing wrong with that. No where is it written that you must seek more responsibility, a higher status, or even more pay. The issue is whether you are satisfied because your needs are genuinely fulfilled, or whether you merely say you are satisfied because you don't think you can do better, even though you secretly wish you could.

If the latter is the case, then the issue may be *self-esteem*. Recent research has shown that it's not just women's salary *expectations* that are lower than men's, but also their *hopes*—that is, the maximum they *wanted* to earn was consistently lower than men's maximums:

> A study which questioned 3,347 undergraduates at Brown University, State University of New York at Stoney Brook, Wellesley College, Barnard College, Dartmouth College, and Princeton University, found that women at these six of the country's most prestigious colleges have lower self-esteem and lower aspirations that men, even though their grades are about the same.
>
> Lois A. Monteiro, project coordinator for the study, said that women are still underestimating themselves. They are much less likely to think they're well prepared for graduate or professional school.
>
> A total of 69 percent of women and 56 percent of men said they prepared assignments before each class; 90 percent of women and 77 percent of men said they took careful notes in class. Yet the study shows 75 percent of the women and 59 percent of the men panicked over an assignment or an exam.
>
> Mrs. Monteiro concludes that women worry more about good grades, underestimate their academic ability by over-preparing for exams, and underestimate their career goals.[3]

The fact is that to receive the same pay, a woman must work nine days for every five days a man works, so both expectations and hopes are realistic in terms of today's world. Many women are still grateful to be working at all, and many find it a minor miracle that they have achieved what they have.

The absolute dollar difference in earnings between male and female workers widens with increasing educational attainments. These comparisons are made between men and women with the same educational experience employed on a full-time basis for the same length of time. When occupational levels and incomes are held constant, turnover rates are not significantly different between the sexes, and in fact, women workers have favorable attendance records compared to men. Yet women do not seem to be less job-satisfied than men. Is, then, the lack of self-esteem really the reason for the low advancement rate for women? In an interesting study, representative of the entire U.S. labor force, it was found that although 95 percent of female workers are discriminated against in pay, only 8 percent actually perceived discrimination. Thus it seems that women are simply unaware of the intense discrimination from which they suffer.[4] In other words, the problem is not just of self-esteem but also of ignorance.

What can you do? The first thing is to *inform* yourself of the going salaries in your place of work. You will not be respected if you accept a lower salary than your colleagues. Also check general benefits and perqs (the benefits attached to specific titles and ranks, such as an expense account or a company car). Be sure that you get no less than the men who are at your same level. Be *needed*. Become an expert at something specific or try to be the only person in the office who has access to particular information or has a specific skill. If your demands are not being met, be *persistent*. Even though my classes are closed to any more students beyond a certain number, the individual student who badgers me hard and long enough will most likely get in. This is the experience of other professors, too. Do wear *down* your bosses, but don't wear them *out*.

It is important to know the company's objectives. Then you can work actively toward these objectives thus proving your loyalty. Bosses are jealous people. They want the respect of their employees and as much devotion to work as the employees might show to their own families. This attachment to the company is often an expectation. ''We're like a big family here'' is a statement meant to compete with the loyalty an employee may feel toward her own family. This expectation is usually harder on women than on men. The extra hours spent at work, which are part of the tangible evidence of commitment, are not as easily given up by women with families. (A recent trend is beginning to make an impact: Many men are less willing to give up family time for work.) It is important to be seen as reliable and as credible. This is a critical attribute of people on the move. To achieve this reputation, make no promises you cannot fulfill and always carry out all that you promise. Level with others, even if it is disagreeable. Be scrupulously honest in all dealings with colleagues and bosses. Look for all the occasions you can find to practice problem solving, decision making, and planning, and attend meetings and represent your company whenever possible.

SETTING PRECEDENTS

There are prices to pay for being the first. The women who moved in where none existed before are like the head goose in a flying formation. The first one gets the full force of the wind but diminishes the impact of it upon the others. After a while, the head goose flies to the end of the formation for a rest and another one takes over. We do not have that opportunity for rest; we have the wind blowing in our faces most of the time. We are put down, dismissed, passed over; we are discounted if we don't, we are sanctioned if we do. Having to choose between indifference and hostility can become debilitating. Constant watchfulness in order to ''do it just right'' is emotionally draining. We need to guard our emotional reserves, to recharge our depleted energies. Somewhere in our lives we need support, encouragement, understanding, time for ourselves and *fun!* Do we pay a price for all this? Do we become more labile?

Some women will never be first because they have never learned to depersonalize criticism. To them, criticism indicates a conflict between the person criticized and the person criticizing. In reality, it often is simply feedback on performance, not a personal judgment. Since most women have been taught to avoid conflict, facing criticism becomes impossible. Therefore, they play it safe. When you initiate ideas, voice a different opinion, argue a point, stand up for an unpopular position, or take on more responsibility than expected, you're up front and very visible—you risk criticism and conflict. To avoid the bad feelings created by conflict, some women never go beyond the expected. The *Paths to Power* are strewn with conflicts to be confronted and overcome. Each time a conflict is confronted, a risk is taken and the confrontation gets easier. Effective managers are initiators and risk takers. Be your own judge: When the issue is important, take a stand; when it is not worthwhile, don't expend much energy. Recognizing the difference and acting on it comes from both knowledge and knowing.

Trust your gut

Knowing is based on the ability most women have to observe better than men. We subliminally pick up even minimal cues, and discern the patterns thus formed. Those gut feelings are good data for us to trust. We have hunches about when things are right, or when something should work or might not. But if you are shy and doubtful about your ability, simply reading that you *should* take risks, confront, and be visible is not going to help you do it.

In looking over the applications for our M.B.A. program, I noticed that all of the men except one had answered a question about future plans with phrases like "I *plan* to" or "I *will*." Only one male wrote "I *hope* to." Of all the women applicants, one-third said "I hope to." It is safer to hope for something than to promise that you will accomplish it, even if only to yourself. But behind those purposeful statements is *a belief*—a belief that you *can* do it—as opposed to just a hope that it might be possible.

This is probably the critical difference between men and women. Men *will* and women *hope*. Why? It is still a societal expectation. When a man achieves something it is normal, expected. When a woman achieves the same thing, it is extraordinary, unexpected. When a male colleague of mine recently published a book, his family and friends thought it was about time. Now that I am publishing this one, my family and friends find it absolutely wondrous. My colleague had planned to; I had hoped to. In a way, of course, it is very nice to have people think I'm doing something extraordinary; on the other hand, it is also somehow demeaning because of the message of lower expectations.

We know that successful career patterns are often correlated with the initial expectations the person has about her own future. If you expect to do well, you exhibit this expectation to others. Your belief in yourself is contagious. Others will catch it and believe in you and your potential. This is a self-fulfilling prophecy.

How do you get to believe in yourself? Let us try a simple exercise. Would you hire yourself for the job that you have or the job you wish to have? List all the reasons why you *would*. If these reasons are true, then at least in your head you know your own worth. Now you must transfer that realization from your head to your solar plexus and transform it into a pervasive feeling of self-confidence and high self-esteem. In turn, that feeling must be transformed into a *visible attitude* that you are a competent person and an asset to the organization. It is critical for you to keep working at achieving the feelings of self-worth.

What may help you to appear more self-confident than you really feel, is to pretend you're another ''you''—a self-confident ''you''—and then play that part as if you were on a stage. If you can do this, you should eventually be able to integrate the feeling enough to own it as part of you. At that point, you won't just be playing a role anymore.

Are you availing yourself of all possible opportunities? Are you utilizing your competencies to your satisfaction? Are you living according to a value system that makes sense to you? Are you fulfilling your true obligations and needs, not those imposed by socialization? If you are even *attempting* all this, you are already a most worthwhile person. It is next to impossible to achieve everything, but it's important to be aware of your goal. We cannot be Superwomen, but we are women, always searching, always trying, always caring.

As you begin to get your bearings and the landmarks become more familiar, you begin to feel more at home with the people around you. You have successfully moved in and are starting to settle down.

NOTES

1. Anne Harlan and Carol Weiss, ''Moving Up, Women in Managerial Careers,'' on-going research (Wellesley, MA: Wellesley College Center for Research on Women).
2. Fritz Steele and Stephen Jenks, *The Feel of the Work Place* (Reading, MA: Addison-Wesley, 1977).
3. ''Women at Top Colleges Show Lower Self-esteem Than Men,'' *Foster's Daily Democrat* (Dover, NH), December 4, 1978, p. 12.
4. Charles Weaver, ''The Irony of the Job Satisfaction of Females,'' *The Personnel Administrator,* May 1979, pp. 70–74.

SELECTED READINGS

Bardwick, Judith, and Donovan, Elizabeth. ''Ambivalence: The Socialization of Women.'' In *Women in Sexist Society,* ed. by Vivian Gornick and Barbara K. Moran. New York: Basic Books, 1971.

Fast, Julius. *Body Language.* New York: M. Evans & Co., 1970.

Giffen, Kim, and Patton, Bobby R. (Eds.). *Basic Readings in Interpersonal Communications.* New York: Harper & Row, 1971.

Mead, Margaret. *Male and Female.* New York: W. H. Morrow, 1949.

Pierce, Carol. ''Women's Personal Issues.'' Unpublished paper (Durham, N.H.: New Dynamics Associates, 1974).

Pierce, Carol, and Eddy, Janice. ''Man/Woman Dynamics—Some Typical Communication Patterns.'' Unpublished paper (Durham, N.H.: New Dynamics Associates, 1974).

Pogrebin, Letty Cottin. *Getting Yours—How to Make the System Work for the Working Woman.* New York: Avon, 1975.

Rohrbaugh, Joanna Bunker. "Femininity on the Line." *Psychology Today,* August 1979.

Shaflin, Alice, and Shaflin, Albert E. *Body Language and the Social Order.* Englewood Cliffs, NJ: Prentice-Hall, 1972.

Woods, Marion M. "What Does It Take for a Woman to Make It in Management?" *Personnel Journal,* 1975.

Chapter 4

GOOD MANAGEMENT POTENTIAL

If I'm assertive
I'm seen as aggressive,
If I'm aggressive I'm a bitch—
I won't be promoted.

Let's try it again:
If I'm nonassertive
I'm seen as a patsy,
If I'm a patsy
I won't be promoted.

Let's try it once more:
If I'm very careful
I can go unnoticed;
If I'm unnoticed,
No one will know
I want to be
promoted.

Any suggestions?

Settling Down

You have lunch often with your co-workers and maybe on occasion with your boss; you are beginning to get a handle on the job. You have placed a photograph or some other personal item on your desk, and you are not the "new girl on the block" any more.

As you figure out how the place works, who does what and how well, who expects what, who rewards whom, what is challenging and exciting, what is boring or a dead end, you will begin to know if you like things as they are or if you want to move ahead. Each organization has its own norms about getting ahead. One of these that you have to identify is how hard and how fast others work. If you work better or produce more than your colleagues, you may gain recognition from your boss, but this kind of norm breaking might also lead to being ostracized by your co-workers. It makes you look good, but the others look bad.

After establishing a initial level of comfort by knowing what the expectations are and how others meet them, you have to decide whether you will opt for good collegial relationships and stay where you are, or whether you will risk experiencing animosity, envy, and even hostility in order to rise above the others. This decision will be based on your career aspirations, but it is not an easy choice, especially since support is so crucial for survival.

BEGINNING THE CLIMB

If you aspire to move up and are currently at an entry-level job—a trainee, a secretary, a sales representative, whatever—set your sights on the next level. One way to find out the steps necessary to get to that next level is to ask people already there to tell you how *they* made it, what steps they took. Most people enjoy talking about themselves and your questions will give them an occasion to do so.

71

Here are some very general points (see also Appendix C, pp. 256–261):

- As soon as you have mastered your job and are beginning to get bored, it's time to look for new challenges. Ask for different assignments; get on work teams; attach yourself to new projects. This will broaden your scope, but you must do it with a narrow focus in mind, that focus being your *next* position.

- Just doing your job well will not get you promoted. Your boss must *notice* that you are doing it well, must *notice* that you are working hard, and—this is a very important point—must *be informed* that you aspire to move on.

- It is important for you to recognize which jobs are in themselves dead ends—in other words, there's no place to go from these jobs, or no one has ever been promoted from them. If you're in one, ask your boss about making a lateral move, a move to get you into a job that *does* have potential for upward movement.

- Think of your position in terms of responsibility. If I asked you right now to describe your job and you answered in terms of tasks you do routinely, you may be stuck. Many women tend to see their jobs that way. Most men, on the other hand, would answer in terms of the responsibilities that they have.

- For many women, admitting that they are ambitious is almost shameful. Ambition smacks of competition, of manipulation, of overtaking others and disregarding their feelings, of pushing yourself up at the expense of everyone else, of being "power hungry." If this is how you feel about ambition, then you will indeed have a difficult time admitting that you are ambitious, and asking your boss to help you strategize in order to advance will be equally hard. However, if you could look at ambition as an *opportunity*—to fulfill your potential, to serve the organization, to manage people in a humane way, to make money, live better, and enjoy life—then perhaps you will feel more 'legitimate' about going to your boss and asking what your next step should be.

- Finally, you must do your best to look and act like a manager—or better still, an executive.

The managerial look

What do you picture in your mind as a woman manager? She can be warm and friendly, but not intimate or cute and flirtatious. She can be caring and supportive, but not a "Jewish mother." She will *not* gossip, dwell on the personal problems of others, or share her own. She will be task-oriented, people-concerned as it relates to the task, and well informed of organizational politics.

Did I just paint a rather stark picture? Now add a neutral-colored suit, understated jewelry, medium heels (unless you don't wobble in high ones), a good-looking briefcase containing a small clutch purse, not too much make-up, and a discrete perfume, and you have a woman who looks like the stereotype of an executive. Don't snicker at stereotypes—they are very real; you might as well use them to your advantage.

To become an executive, you must look like what others believe a woman executive *should* look like; once you're there, you can do as you please. How do you know what that look is, what to wear? The criteria are simple: Always be understated, never flashy. Underplay hips and bosom, don't accentuate curves. Wear clothes that allow you to cross your legs comfortably. Your clothes don't have to be expensive, but they must *look* it! No shiny knits; the materials must look like cotton, silk, linen, or wool even if it is polyester. Pant suits are permissible only when they are very elegant. But if all the secretaries wear them, don't. In other words, notice what the women at higher levels wear and dress accordingly.

Perhaps I am a little conservative, but it is better to err on the side of formality than on the side of camaraderie if you want to be taken seriously. With time and more acceptance, you will be able to change to a more relaxed posture.

You do your job well, you look every inch a vice president, and you've decided that you'd like to move on to bigger challenges—and your boss knows this. What else should you do? There are four strategic moves to make in order to move up. The first is to *understand the controlling forces* in your—in *any*—organization. The second is to *master specific and general skills*. The third is to *use politics*. The fourth is to *overcome the inhibitory factors of non-assertion*.

UNDERSTAND THE CONTROLLING FORCES

The path that leads away from dependence towards more autonomy has two major external roadblocks. The first is discriminatory company policies or, at the least, generally unsympathetic attitudes toward women. The second is a superior who either purposely blocks your advancement or does nothing to promote it. But it is not necessarily just one individual who is responsible for the difficulties you may encounter at work, but a body of people—a nameless, faceless "they," the powers that be—that is only dimly perceived and only superficially dealt with. The system, institution, the organization, the board of directors, the trustees, the governing bodies—"they" decide the various *roles,* distribute the *tasks,* allocate the *resources.* If you can understand roles, tasks, and resources in terms of expectations (theirs and yours), learning what is negotiable in one or more of these areas, you then have a chance to exert control. Role deals with who you are, task with what you do, and resources with what you have.

Role

How do you recognize who has what role? Role is usually determined by formal title, by the responsibilities attached to that title, and by the behaviors thus exhibited. However, a fancy title without responsibilities, adequate pay, or an appropriate office will not provide influence. The way that a specific role is performed differs with the people who occupy that role as well as with the situation. Role is also the part you may take in interpersonal and organizational dynamics, such as the confronter or the harmonizer. For instance, the role of a chairperson can be exercised democratically, seeing to it that all members have a say, or manipulatively, seeing to it that only those members with opinions similar to your own are recognized to speak. The role of secretary may go anywhere from typing, filing, and sewing on the boss' loose buttons to attending meetings, writing reports, and making important decisions. Roles always have two aspects: the expectations of others and your own expectations.

The importance of others' expectations is that they color their perceptions of the performance. As an example, take the expectation that a manager will maintain a high level of productivity in a unit. It might then be concluded that the manager's role is to exert control over the people in the unit, to put on enough pressure to reach the expected productivity level. If the manager is a tough, directive, demanding, and critical man, he is within the traditional role expectations and will be perceived as appropriate. However, if a woman manager does exactly the same thing, the expectations may be that she should accomplish the same objective by being more tolerant, more understanding, more sympathetic, and less critical. If she performs like the man, she is seen as role-inappropriate and is violating expectations. On the other hand, if she behaves according to female role expectations, she might then be seen as weak, soft, too permissive, and not having a tight enough grip on her unit.

Because of these different role expectations, women are often seen as not being able to perform well the traditionally male tasks; they are also only infrequently given the opportunity to prove the contrary. Instead, women are given less prominent roles with fewer responsibilities, and narrower chances for advancement. With the lesser roles go fewer opportunities for organizational contact, both horizontally and vertically, and fewer contacts with people outside the organization. Without the chance to be visible or the room to be innovative, women become role-bound and stuck in their positions.

In our culture, nurturing and caring and all maintenance functions (cooking, cleaning, washing, and taking care of the young, the old, and the ill) are generally performed by the subordinate class and is therefore devalued. Women and minorities constitute the lower class and do the menial work; the dominant class of white males pay for these services to be performed.[1] The sad thing about all this is that the women who wish to get out of the subordinate role in order to be part of the dominant class believe they must first sacrifice an important part of themselves, the part that supports, gives, loves. This need not be so. You can expand your range to take on *all* roles appropriate to a situation, to be both tough at times and soft at others, confronting here but supporting there.

What strategy should you pursue in terms of role? If assigned a lesser role, start to perform at the next level so that you take on functions beyond your job description. You have to act like a vice president before you can become one! You could be seen as pushy and threatening, but if you don't push, you won't be considered management material. You must tread a fine line between the two, asking questions and getting feedback along the way from colleagues and bosses as to how you are doing.

Task

Task deals with what is to be done. This often, but not necessarily, depends on role. What you actually do and how you do it is of critical importance. For example, when asked to write a report, you not only write the report as requested, but also add a section of your own suggestions or ideas and send it to several people who are in key positions. If you see something in the office that needs to be done differently, initiate not only the idea of changing, but also the changes themselves, asking other people for help. By being in charge of several people, you then assume a role of more influence, and therefore more control and more power. Traditionally, women have been given tasks with little responsibility but a lot of accountability. It is important to see to it that the responsibility and accountability are more balanced. It is unfair to be accountable for something but not be totally responsible for it. You must see to it that you neither get blamed if the task goes wrong nor miss the benefits of a successful outcome. Every major success should permit you to ask for more responsibilities. Of course, before any new venture there is the fear of failing to be overcome, but the old adage, "nothing ventured, nothing gained," should be the motto of would-be women managers.

Resources

There is a third dimension of paramount importance—the use of resources. This encompasses such items as money, equipment, people, services, information, territory (both the physical space you occupy and the size of the territory you control or manage), time, and finally your own knowledge, competencies, and skills.

Money deals with the budget allocated to you or to your department. The more money you have, the more you can expand, innovate, get more or better people to work for you, have more or better equipment (such as special machinery), get improved inter-office mail services, use outside services such as advertising agencies. These are rather evident. What I want to deal with here is the resources that affect women in a special way: *information, territory,* and *time.*

Decisions are made based on information. Access to information is not only through formal channels, such as written reports or exchanges at meetings, but also through the informal network of an organization. It is shared during the coffee break, in the bathroom, in the car pool, or after hours over a drink. Many decisions are made, deals confirmed, and promotions promised via this informal exchange. Yet many women often find it difficult to get included in that network. It is important for you to make a concerted effort to see your colleagues outside of working hours or during lunch—at the least, spend some time talking with them about both office and non-office business. There is no need to *drink* with them if you do not drink, but there is a need to go out with them. Don't be the only one never asked. How do you get asked? You don't—*you* do the asking—*you* initiate the interest, the contacts. No, it's not easy, and you may shortly have to get home to feed the kids anyway, but *do it*. It's tough for women who have responsibilities at home to be part of the informal outings that often also take place on weekends or in the evenings. However, you must be included in this informal network to advance. It alone gives access to one of the most valuable resources of all: information.

At some office functions, your male colleagues will bring their wives. If you have a husband or a male friend, he will be in a very difficult role: He cannot take part in the office talk nor is he one of the women. Yet it might be important for you to share a part of your personal life at least once in a while, if others have done so. I have also found that spending an evening with my all-male colleagues who have their wives along can be equally difficult. I am torn between wanting to discuss courses, students, the university, and faculty gossip with my colleagues and wanting to sit with the women and talk of things that interest me just as much. I end up going from one to the other, catching only bits and pieces of conversations and information.

Territory is important, in terms of both type and location. This refers to your particular status in the company. Those who manage the largest departments, the most important projects, the most productive sales group naturally have more influence power. The other thing they have is an appropriate office—personal territory.

We all know that the corner office has the higher status. Several new office buildings in Toronto, Canada have been built in the shape of a cross to provide space for eight corner offices. When I became the first-year MBA program coordinator, I moved from a small office that I shared to a large office that was all mine. I still can't get over the different feeling it gave me about myself—did I only imagine that my peers and my students respected me more? The larger office also allowed me to call meetings with my colleagues there; I was on home ground and therefore legitimately in charge. The person who calls the meetings and sets the agenda is in control. That person also makes sure that whatever she believes needs doing gets done, provided it meets with the goals of the organization. It was from my larger office that many of the projects I initiated got carried through. Invite people often into your office and offer coffee, if possible. In this way, your office will become a place where information is shared and decisions are made. In addition, the closer you are to the bosses, the more opportunities for interaction, the more visibility, the more chances for advancement. I would rather take a smaller office close to my superior than a larger one farther away.

Time is a third important resource, but a more illusive (and sometimes elusive) one. The responsibility of a task is often measured by the amount of time taken to perform it: The longer the time needed, the more important the task. Projects that take only a few hours or even a few days are not as critical to the organization as those that take weeks, months, or years. Only the people at the very top make the decisions whose results are projected in years, such as general policy. In other words, the shorter the time allotted to something you do, the less it will benefit you.

One strategy for slipping out of this mold is to try to delegate all tasks that don't take a lot of time. There is one exception: crisis management. When there is an emergency, someone has to make a quick decision. Let that someone be you, whenever possible. Decision making is an opportunity seldom given to women. Taking a chance will give you a chance to prove yourself. Time also deals with boundaries, that is, when you are to start a project and when you are to end it. The more control you have over those boundaries, the more autonomy you have and the more creative you can become. It is only when you do tasks that are nonroutine and innovative that you will have a chance of gaining any recognition and respect for your competence.

Another aspect of time is the difference between exempt and nonexempt employee categories. Nonexempt people have to put in a certain amount of hours each week and are paid overtime for any hours worked beyond that limit. Generally, these are lower-level jobs. Positions with more responsibility are exempt from overtime; therefore, the people in those positions may pursue a different schedule. Managers stay later, take longer lunches with customers, travel, and are generally independent of anyone else's schedule except as appropriate to the task at hand. Your time becomes more and more your own as you move up in the organization.

Role, task, and resources often go together. If you presently have none, look at all three dimensions and decide which you ought to pursue first. The nameless, faceless "they" who gave you an inconsequential role, assigned you an insignificant task, and allotted you limited resources are in for a surprise. Most women are not used to aspiring to more or better pieces of the pie, perhaps because they did not know how to go about evaluating these three dimensions. It is always easier to deal with small pieces than tackle an abstract totality. Focus on one dimension at a time, in order to use your energy productively instead of diffusing it in frustration.

MASTER SPECIFIC AND GENERAL SKILLS

Each job has its own particular knowledge base and skills. You need to be competent at whatever job you have been hired to do. With some hard work you should become really good. However, do not lose sight of how your competence fulfills the larger organizational objectives. You need to prioritize: Don't do only what is most meaningful to you in the present, but also what will be significant for your department (or company) in the future. You must go beyond competent to effective. Effectiveness is good use of competency.

Besides the specific skills you need for performing competently, you need to master the general skills that effective managers practice on a routine basis. I call them the Four Uncommon Skills because even though they are commonly practiced, they are *un*commonly practiced *well*.

The four uncommon skills

Reading, writing, counting, speaking! You may be good at one or two of them, but seldom is anyone really accomplished in all four. Yet an effective professional ought to be at least competent in them all. You should be able to figure fractions and percentages and to look at charts, computer printouts, and statistics without going blank. If you cannot speed-read, you should at least know how to skim over reading material. Your writing skills need to be good enough for you to say what you mean using precise and concise language. Finally, speaking in public gives you the opportunity to be most visible and to make your presence known. You must feel comfortable in this position.

Reading   Reading for information is a primary way to be up-to-date. To advance in your company you *must* be knowledgeable about the current development of your profession; if you aspire to be a manager read the *Wall Street Journal, Business Week,* and the trade journals of your specific interest area.

You also should be knowledgeable about the general economic, social, and political scenes in the world. The world is a vast system; anything that happens in one corner of it in some fashion influences the rest. To be up-to-date on this information requires knowing recent Supreme Court rulings, understanding the stock market, knowing the platform of local state and national politicians, and being aware of demographic changes, especially those you feel may in some way affect decision making at the highest level in your organization. If you add to this list your own particular interests, such as the theater, sports, or fiction, you could end up spending all your time just keeping abreast.

However, only a minimal amount of information is necessary to function intelligently, so you simply need to read fast. Try scanning, which means glancing at a page in order to know whether there is any valid information in it for you, or skimming, which means reading through very fast to pick up key concepts. Otherwise, I would suggest you take a speed-reading course. I did, and I am certainly reading faster for information than I did before. For instance, I peruse all information pertaining to women's issues and have thus acquired a wealth of data.

It is possible to narrow your focus and read only trade journals and professional magazines if that is all you are able to do in order to keep up. However, you are likely to find that your climb up the ladder will eventually be forestalled by such a narrow outlook. The people at the top *do* keep current on national and international news and events. It can be a great pleasure and an important enrichment to a broadened point of view, but not at the cost of turning you into Superwoman by forcing yourself to "know everything." Priority and choice are important. Be aware of what you should read in order to acquire the knowledge you require.

Writing   When you write a memo, a letter, a report, or a speech, you are *transmitting information.* Therefore, you must be precise, so the message is understood, and brief, so the message doesn't get thrown away or put aside for later reading. What helps me clarify my message is to first put down in a stream-of-consciousness format all thoughts that come to mind—like brainstorming with myself. Then I can organize them efficiently.

You must point out the *significance* of your information for your readers, unless it is self-evident or was previously requested. Think about how the information affects the people who will receive it. This will add personal significance to your message. Your ideas should flow in a logical sequence. If you have written a fairly long piece, check the order by reading it back and making an outline from it. If you cannot, the piece is not well organized. Try to anticipate any questions the reader might ask and answer them in a clear and logical way. It's wise to introduce only one thought per paragraph, and major sentences running longer than four lines will probably lose the reader.

Try to use the active rather than the passive voice. It is much more impactful to say "I did" than "It was done." For instance, instead of writing "It is thought that,'' use the more active ''The department thinks that.'' Don't write ''The issue that was to be looked into,'' write, ''The issue I looked into.''

Writing can clarify or confuse, simplify or complicate, inform or distort, intrigue or bore.[2] Pay attention to your writing—it expresses you as much as your speaking voice and your nonverbal language.

Counting   If numbers frighten you, if you skip over statistical charts because you feel you cannot understand them, if you have trouble reading a computer printout, if financial statements are Greek to you, if you cannot figure out the profit and loss from an annual report, you need help—especially if you also aspire to move up in the organization. Either spend some time with a friend who can help you with whatever gives you difficulty or take a basic accounting course.

Most of the major managerial decisions include budgetary concerns: What are the resources in terms of monies available? What are the costs in terms of monies spent? For many businesses, basic arithmetic may be sufficient, but it is interesting to note that most of the people who made it to the top in the past few years have come through the financial route, be it vice president in charge of finance or comptroller. In the 1950s, when there was a shortage of goods, the route was through production. In the 1960s, when there was competition for the distribution of goods, it was through marketing. In the 1970s, the need to control costs and to acquire capital made it possible for the financial expert to come up through the ranks. This is still true today, so the need to be at ease with numbers remains. Some people predict that planners will be the executives of the future, which will involve financial expertise as well.

Do not acknowledge that voice within you that says, ''I can't do math.'' A great deal of recent research points to the fact that women have been ''socialized'' to avoid math while men have been socialized to persevere in it. If you are math-anxious there is nothing wrong with you, nor with your mind. You merely lack a certain skill, which can be learned, provided the atmosphere you learn it in is supportive. There is no such thing as a person who does not have a good mathematical mind. You would never say that someone does not have a good historical mind, or a good geographical mind. If you feel handicapped by your lack of comfort with math, you *can* do something about it, and I would urge you to find out what courses are being offered in your community.

Some universities offer special courses for the math-anxious. I was a math avoider all my life and am so math-anxious that I decided to chair a committee to study math anxiety on our campus. We uncovered that of a random sample of 1,000 faculty and administrators to whom we sent a questionnaire, 100 responded that they would be willing to take a special workshop dealing with the subject. If you are too anxious to take a course, face that fear and go. If you do not, your chances for advancement will be very slim.

Speaking   Speaking is the most visible of the four uncommon skills. You may not be well read, you may not know how to count, you may write poorly, but as soon as you open your mouth people get an impression of you based both on the content of your message and on the way you deliver it. If you attend a lot of formal meetings

you should know the *Roberts Rules of Order*. You can find them in the College Outline Series.

Unfortunately, many of us tend to put ourselves down, both by the words we use
and by smiling a lot. We put ourselves down by using qualifying statements that
diminish what we say. For instance, we start sentences with, "May I say something? . . ." or "I don't want to interrupt, but, . . ." or "I may be wrong,
but . . ." or "I'm not sure about this, however, . . ." We also end too many sentences with a question mark and use phraseology such as "I kind of," or "I sort of."
The use of superlatives also takes away from the impact of our statements. Using
"terrific," "extraordinary," and "fantastic" too much makes you *less* credible.

Watch how frequently women smile or even laugh when they say something that
is neither pleasant nor funny. It is an indication of discomfort at being visible. Avoiding qualifying statements and not smiling aren't the only things that will make you
be heard. Eye contact, the steadiness of your voice, and its volume also have an impact. Given the choice, you are better off speaking too loud than too soft. I use a specific strategy when I am in a group where many people are vying for air time. When I
speak after someone else has finished, I make a mental note of how that person
spoke. If it was in slow, measured tones, then I speak faster. If that person spoke
quickly, I speak slower. If that person was high pitched, I pitch my voice lower. In
other words, I try to make the difference between my style and that of the other person just jarring enough to make people pay attention.

Here are some other tips I have found useful. I never make a speech without being *prepared*. And unless I know my subject really cold, I always have cards with my
topic headings, just in case I go blank. I never have yet, but there is always that possibility lurking in the background. When I first began speaking in public I used to
practice in front of a mirror or in front of friends. When my daughter was a teenager
she had to hear all my early lectures before I would go in front of a classroom. That
was very helpful.

It is okay to be keyed up; the audience very seldom knows it. If panic sets in, just
read off your cards, but *stick with it* and eventually the fear goes away. The more you
do it, the less fearful you will be. I always used to have terrible stomachaches before I
spoke and still do when unfamiliar situations arise. What helps is trying to change
postures and positions. If I am standing I tend to walk up and down, but I have to
watch not to overdo it. However, it is better to pace than to stand rigidly in one spot.
Also, if I am very nervous I tend to talk fast, in a high-pitched voice, so I make myself
talk slower and lower my pitch. If I am sitting down I move back into the chair so my
back pushes against the seat instead of leaning nervously forward.

Alcohol and tranquilizers do not help because they make me feel *less* focused
when I'm attempting to be more so. Rehearsing into a tape recorder helps, even
though the voice never sounds like your own. When I have finished talking and there
is to be a question-and-answer period, I'm always ready with the following in case
there are no questions from the audience: "If there are no questions, then I shall
close with," and I make my prepared statement. This way, I do not end with a failure
to elicit questions, but with a closing statement of my own. I've also found that it is
useless to make more than three major points in any one speech. People tend to forget, become confused, and remember nothing if too many are discussed.

Women must work twice as hard as men to gain the floor and the attention. Two California sociologists, Candace West and Donald Zimmerman, found that men interrupt women in conversation far more often then they interrupt other men and more frequently than women interrupt anybody. Researchers argue that interruptions reflect and assert power differences. The more powerful conversational partners interrupt the less powerful ones more frequently.[3]

Pamela Fishman recorded hours of conversation and found that topics introduced by men succeeded (that is, were developed further) 96 percent of the time, whereas topics brought up by women succeeded only 36 percent of the time, even though the women worked much harder at initiating the conversation.[4]

It is interesting to note here that what is called "women's language" (using qualifying statements or superlatives and ending opinions with question marks) is not used only by women, but by most people in low-power positions. Researchers studied a police station and noted that this kind of language was used more by clients, whether male or female. The "male" type of language was used more by the police officers, also whether male or female. This confirms once more that, in general, what we consider to be "femaleness" is powerlessness.

De-sexing language

As women, it is to our advantage to de-sex language. We often fall into the trap of describing women in terms of how they look and men in terms of what they do. For instance, we might say "Bill is the accountant" but "Jane is the attractive redhead" or "Jane is the one in the lovely two-piece dress." Job titles are frequently sexist: businessman, mailman, salesman, fireman, policeman. These could all be translated into business executive, mail carrier, salesperson, fire fighter, and police officer. More and more women correct men who call them "girls"; the men certainly would not call men from a similar age group "boys."

I don't correct everybody *all* the time; it can become tedious. But I do it enough to make people aware that their speech reflects unconscious sexism, or at least unexamined stereotypes. When I first went into my publisher's business office, I found in the lobby a board with the names of three visitors: two gentlemen—one doctor so-and-so and a professor so-and-so—and "Ms. Josefowitz." I am both a doctor and a professor. I complained and was met by shoulder shrugs. How much do I push? I didn't then. But if it happens next time, I will mention it again.

THE PLOYS OF THE POWERLESS

You are in the most dependent power position; you have no direct influence over others, most others have influence over you, and you have little control over your own life. Powerlessness tends to make people feel helpless, hopeless, frustrated, or hostile. When people feel powerless they often become either depressed or angry. Depression is turned inward into poor performance and ill health. Anger is turned outward into acting-out behavior. Both depression and anger are often translated into poor attitude at work. Women at lower levels lean more toward depression as an expression of powerlessness. What can you do to change this? What power can you exert, what influences can you muster to make life at the bottom more palatable?

The powerless are not without power. If we see powerlessness as the dependence on something we want or need from someone else, then there are six possible strategies:

Influence through reciprocity This strategy makes the person on whom you are dependent also dependent on you, whether for favors, for approval, or for admiration. Reciprocity is based on indebtedness. If you have done someone a favor, that person owes you one. If someone has done something for you, be sure to acknowledge it, and if you can, do something in return.

One of the strategies of the powerless is to identify what is that *they* have that the powerful may want. What could a man want from a woman? How about warmth, nurturance, approval, sympathy, affection? How about special skills or knowledge, intuition, wisdom, creativity, responsibility, initiative, stick-to-itness, strength, understanding, intelligence, problem-solving ability, fairness, good judgment? I could go on and name every attribute in the book. Which ones do *you* own? Which ones do you recognize as yours? These are your resources, to be given or withheld.

Freedom through alternatives This strategy will free you from dependence on a single person. If you spread the means of satisfying your needs by being dependent on several people, no one person will have total power over you. Even though you have one boss, you may be able to find other people from whom you can gain recognition for your work, if she or he does not, and who will give you support, advice, encouragement, or whatever it is that you need and would like to receive from your boss but cannot obtain. You then have other people to fall back on.

Pressure through numbers This is the formation of support groups, or coalitions, as a strategy to confront the person or persons in power. The visibility and strength of the group voice is exemplified by unions. The real power of the powerless is in numbers! The more women there are, the louder their voices, the more there can be honest talk, direct confrontations, demand for explanations. This dispenses with the individual's indirect manipulation and ploys.

Visibility of your own knowledge and skills It is not enough, however, merely to know or to perform. You must also make this knowledge heard and the performance seen. This expertise may take the form of being the only one to know how things are filed or being the most skilled in a specific field. Especially at entry-level positions, it is more important to establish yourself as *the* expert in a narrow area than to know a little about a lot. As you move up you will also need to do the latter, but whatever you do, what matters is that you are noticed and recognized by others as possessing that particular knowledge or skill.

Control of resources Remember that whoever *controls* resources has power. One of the resources the powerless frequently have is *information*. Office gossip can be a powerful resource. Gather all the information you can, but be careful *not* to malign anyone, for you will be quoted.

Access to influential people Take any pretext to talk to influential people—ask them questions, offer to help, send them articles that might interest them. This could result in three things. The first is possible access to the resources these people control. The second is being attributed power by others through being seen with powerful people. The third is "familiarity breeds promotion."

Let us take the example of a secretary. She is controlled by her boss, but she may have access to information he wants, and she can therefore use their working relationship to influence him. She can also either facilitate or block others' access to him. This is sometimes known as the power behind the throne.

Contingency style of subordination

The powerless need to use whatever tools are at hand and to expand their range of strategies. However, as long as you are in a powerless position, and until your strategies begin to work for you, use a contingency style of subordination. In other words, your subordination will depend on your boss's personality and on the situation. It will at least make your time in the lower-level positions more bearable. You need to diagnose your boss in order to learn complimentary attitudes: With an autocratic boss you submit; with a laissez-faire one you take over; with a democratic one you problem solve together; with one who seeks information you provide it; and with one who likes to give information, you request it and appreciate it.

This type of manipulation may go against your grain. Playing politics does not suit everyone. If it makes you very uncomfortable, you may decide to sit it out and see what happens. It is likely, however, that not much will happen. It is important for you to understand that there are costs and benefits to any action you undertake or decide *not* to undertake. In other words, "inaction" is as much a decision and an act as "action."

BECOMING MORE ASSERTIVE

Assertion. Is it a skill? Is it an attitude? Are we born assertive? Do we learn it as children? Can we acquire it as adults? Most of the literature on assertion assumes that people are unassertive because they don't know *how* to be assertive; they lack the skill. In other words, the assumption is that we can be *taught* to be assertive with practice. Most assertiveness workshops use role play on the premise that if you practice assertion in simulated situations, the behaviors will be transferable to real events. Of course, practice can help people discover the various ways of approaching a topic or person, but practice will not help the people who are shy or afraid to act.

So the question to raise is what makes some people too shy or too afraid to get their needs met?

Current research indicates that there is an *inhibiting* factor that prevents some people from being assertive.* Shy or fearful people do not feel entitled to express their thoughts and feelings, make requests, or refuse the requests of others. They

---

*My daughter, Nina Josefowitz, is a doctoral candidate in the Department of Applied Psychology, University of Toronto. Her dissertation is on nonassertion and was very helpful in formulating the concepts in this chapter.

have low self-esteem. They do not think that they are good enough, smart enough, attractive enough, whatever, and feel, therefore that they have no rights and they perceive others as granting them no rights.

Inhibition results in the inability to speak out or act on your own behalf, on behalf of others, on behalf of an idea or value system.

Let us go a step further. On what is inhibition based? It is based on the premise that attempts to be assertive will meet with a negative response, and that this negative response matters—in other words, you care. If you predict a negative response but feel it does not matter, you won't be inhibited. But when it *does* matter to you, the predicted negative response can so influence your behavior that you are prevented from being assertive.

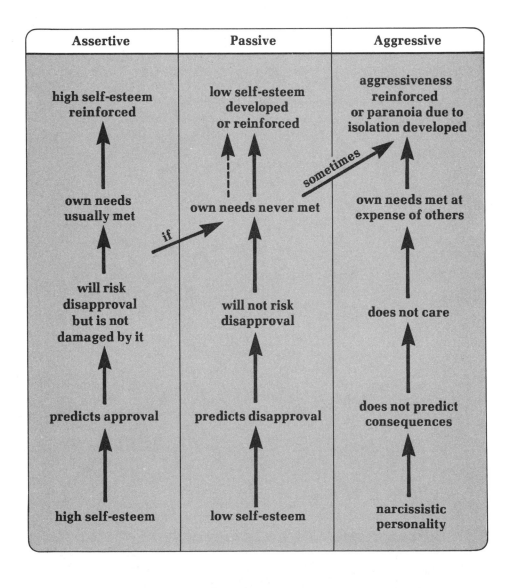

Here's an example: You want to borrow my sweater and I really don't want to lend it to you. I predict that you will be angry if I refuse. If I care that you will direct your anger at me, it can control my behavior and I will lend you the sweater. If dealing with your anger does not bother me, I will be able to refuse, which is what I want to do.

There may be both rational and irrational components to inhibition. You may be correct or incorrect in assuming that a particular assertive stance on your part will result in negative consequences for you. There are a couple of ways to check reality. One is to ask others. If no one else would be inhibited in your place, then you can question the reality of your predictions of a negative response. The other is to recall the ways your parents controlled you as a child. Did they show their disapproval by anger, by tears, by indifference? Does the anger, the tears, the indifference of others control you now? When you feel inhibited from being assertive, visualize the type of negative response you predict your behavior will elicit. Does it remind you of an earlier parental reaction? If you had a father who controlled you by anger or even just by its threat ("Wait till Daddy comes home! You're gonna get it!"), do you predict angry responses to your attempts at meeting your needs? If your mother cried or became ill when you misbehaved ("You'll be the death of me!"), do you now most fear hurting people?

Since the inhibitory factor in assertion is the displacement of early childhood socialization patterns into the present, determine which people most inhibit your assertiveness and try to identify what it is in them that reminds you of your parents. It may be a look, a walk, a voice, an expression, an attitude. In psychoanalytic terms, this is called *transference:* You are transfering onto a boss, a colleague, or a subordinate something that once belonged in a parent and that used to inhibit you. Once you are able to identify this element, ask yourself, "Do I choose to be inhibited by this person who is not my parent?" Chances are that similar attributes in a variety of people will trigger this inhibition in you. I have found that very distant men remind me of my father, who was always quite formal, and therefore withdrawal can control me.

Once you have identified the pattern, chances are that whenever you meet it in others it will *still* influence you, even though you are an adult woman. The only way to deal with transference is to keep looking for the pattern so that you can identify it and thus render your unconscious conscious and therefore more under your control.

STRATEGIES FOR ASSERTION

Assertion at the place of work takes the form of *expressing, requesting,* or *refusing* around the three categories of *role, task,* and *resources.* Each *role* has a different level of influence, a different level of power. If you want more power, do you feel you have the right to it? And, if you do, how do you convey that impression? Assertion in role is claiming a higher or better position. It is still a wonder to me how different I felt when I became an assistant professor after having been a lecturer. I really think I walked around a couple of inches taller; I certainly felt better about everything I was doing.

Assertion in *task* isn't just the competence you feel in performing well. Assertion in task is assuming more responsibilities. I know that, for myself, I always seem to be stretching just beyond my comfort, which means never doing it perfectly, but being challenged by that swim upstream, although I do gulp quite a bit of water.

Assertion in terms of *resources* can be asking for more secretarial help or attempting to enlarge your territory. The difference between sharing a large room with dozens of other people, with glass partitions and no privacy, and having your own office with a door will naturally make you feel better about yourself and therefore able to be more assertive. I believe the more you *own* in terms of territory, the more you feel assertive about expanding it. Just as money seems to make more money, power seems to give more power, and assertion permits more assertion.

Where do you start? How do you practice?

For the purposes of this book, I would like to stay with assertion in the place of work, although it applies equally well in the home. Assertion can be seen as going progressively from (1) *expressing* personal thoughts and/or feelings, which includes initiating, deciding, and creating; to (2) *requesting,* demanding, insisting, and confronting; to (3) *refusing,* rejecting, criticizing, dismissing.

Expressing

Expressing our thoughts and feelings, even though it is the least threatening of those three categories, does make us vulnerable by becoming more visible. If the thought is dumb or the feeling inappropriate, the risk is that of being ridiculed, shamed, or seen as stupid. In other words, if you fear expressing your thoughts and feelings it is because you predict a negative response. What you need to do is question who might give you this response and what shape it would take. Then you could fantasize what it would look like and see how you could deal with it. Possibly it would not be quite as catastrophic if you first played out the scenario in your head. What is the worst that can happen? What would you do if you received a negative response? a positive one?

Initiating, being the first to speak up or to propose, is a higher-risk category because we are treading on new ground. Frequently, just being a woman in an organization is already doing something new. Being either a token woman or part of a small minority gives perhaps more visibility than many of us are comfortable with, and visibility promotes "pot shots."

When there were only two or three women at most at our faculty meetings of around thirty men, every time I spoke up it was with a pounding heart. I felt I was speaking for all women and thought special attention was focused on me. It was most disagreeable and made speaking out difficult. Only doing it often enough stopped it from being unusual. The special attention being paid to me as a *woman* changed into more appropriate attention being paid to my *statements.*

The person who stands out, who speaks out, who looks or acts differently is the person who will be noticed, talked about, praised, or criticized. There are those who initiate, who act upon their environment, and those who respond, who are acted upon. There is no single "right" mode for all situations, but if you do not tend to ever initiate anything, what is your fear? An interesting fact about initiation is its nonverbal aspect. Initiating a gesture towards another person is often an act of dominance. The boss will put a hand on a secretary's shoulder in greeting, but not the other way around. A teacher will touch a student, not vice versa. Doctors initiate touch with patients, ministers with parishioners, police officers with accused persons. A male will put his arm around a female colleague easier than a woman will do the same to a man. Gesture initiation is a dominant/subordinate cultural pattern. One of my doctoral students, in a fascinating research, found that members of *both* sexes are perceived as more powerful and intelligent, more masculine and less polite, when they assume the role of toucher rather than receiver.[5] She was interested in seeing whether women were in any way penalized for initiating touch towards a male and found that to the contrary, there is no penalty for sex violations of this type. In fact, these women were perceived as powerful (attributions of power being dominance and forcefulness) yet attractive (attributions of interpersonal attractiveness being warmth and likability).

Decision making is a form of assertion. Whenever you make a decision you state that you have enough information to take the responsibility for a course of action. Taking responsibility is an empowering stance. Being creative is also being assertive: You express a part of yourself, you offer it to the scrutiny of others as a valid expression of yourself. Creativity, decision making, initiating, and expressing are all risk-taking activities.

Requesting

At a higher risk level is the next category, requesting, which can be demanding, insisting, and confronting. What are your needs and rights? Can you request that your needs be met? that your rights be respected? Can you do it in terms of your role, your task, and available resources? Do you need some time off, do you want that report to bear your name, do you want better qualified personnel, do you want to represent your company at that next meeting? Make your desire known—request it, and if your request is legitimate but not heard, demand, insist, and finally confront the other person's deaf ear or refusal. What do you predict will happen? Will that person disapprove of you? What will be the consequences? What did your parents do when you requested something? Do you see a pattern? Do you recognize the old fears operating today? If you do, ask yourself if you can live with the predicted consequences. If yes, go ahead. If no, then protect yourself. You may not be ready as yet and that is okay. Take your time.

Refusing

Refusing, criticizing, rejecting, and dismissing are particularly difficult positions to take for women. They are not "nice!" Yet unless we are able to say "No" when necessary, give negative feedback when constructive, reject when appropriate, and even dismiss when we must, we will never have credibility or legitimacy. We must learn to disapprove, to stand our ground in the face of disagreement, conflict, and adversity.

If you do this, predict the outcome. Will you encounter a negative response? From whom? Does it matter? If yes, then *why* does it matter? If you can answer this and take this awareness with you when you are faced with the opportunity of expressing, requesting, and refusing, you might be able to help yourself deal with the situation in a more assertive manner. The "why" is critical here. If you can identify the parental injunctions that are blocking you in your encounters with others, you will begin to help yourself toward more mature relationships. A word of warning: After years of struggling with shyness, with feelings of inadequacy, with a lack of self-confidence, you won't see dramatic changes after reading this chapter. But if you stay with the new awareness and discuss these issues with family members or friends, I believe improvements can be made. If you agonize somewhat less and hesitate for shorter periods, recognize it as improvement.

Expressing, requesting, and refusing can be applied to the work place in terms of role, task, and resources. You can *express* dissatisfaction with your *role,* as, say, a first-line supervisor. You can *refuse* to help out on a *task* because of your own time pressures. You can *request* larger financial *resources* for a project. Do you think the responses to these will be negative? How much does it really matter, how much do you care? Is it appropriate to the situation? Does it control your behavior?

It may well be that you are better off doing nothing; withdrawing as a strategy may be wise given a particular situation. Assess the "real" potential risks for you, not just your fears of it. Perhaps this chapter will help you differentiate what you carry over from a past and what belongs to today. I believe that the ability to express our feelings and thoughts and to ask for what is right and to refuse what is not makes us people whom others can trust, to whom others can give responsibility and respect. I certainly respect myself better when I am able to stand up for what I believe is right for me, for others, for the organization in which I work, and for the society in which I live. Assert your rights, assume your responsibilities, and reap your rewards!

MOVING OUT

You have mastered the Four Uncommon Skills, you have learned your job well, and you are competent and effective. You dress well, you believe in yourself, you have good relationships with your colleagues, and your boss knows of your ambitions. Still, you are not being promoted. How long should you stay, hoping and waiting, living on promises of "next time there is an opening"? How do you *know* whether to stay or to quit—when to move out because you're not moving on?

Managers are unaccustomed to seeing women executives change jobs and they overreact to even a moderate turnover rate. An executive search firm who queried forty companies about job-hopping found that only two reported a higher rate for women than for men at upper levels.[6] However, at the lower rungs, both men's and women's turnover rates are higher.

You may be in a dead-end job, a job that will get you nowhere in terms of career opportunities, yet you may be perfectly satisfied. It is possible that such a job could provide you with the necessary income, friendly co-workers, and either enough interest and challenge to keep you involved or enough routine so that the job is not demanding, which conserves your energies for personal pursuits.

There is no "should" about being upwardly mobile. If your job has enough meaningful aspects for you, no one but you can say whether you should aspire to something else or be satisfied where you are. However, if you are not satisfied, then you must assess your situation realistically. If you have been on your job two, three, or more years, it is time to start thinking about moving out of it and into something else if you are not moving up from it.

Are you stuck?

What are some of the signals that tell you that you may be stuck?

■ The work is exhausting, yet unchallenging.

■ You are both tired *and* bored.

■ You feel that your co-workers and your boss do not respect you, that you are not a valued member of the team.

■ You have lost respect for your co-workers and/or your boss, and no longer enjoy working with them.

■ You do not believe in your company's products, policies, objectives, and your sense of values is violated.

It is very difficult to work under any one of these conditions. If more than one is present, you really should move on.

Your first attempt should be to get promoted within your department. If you see others being promoted over you whom you consider equally or less qualified, talk to your boss about it and try to find out the reasons. If you are not given increased responsibility as you deserve it, then you are not being recognized. If you are given no opportunities for training but other people are, you must make your demands clear. If there is no chance in your job to gain any visibility, the prognosis for the future is not good.

If promotion within your department is not possible, then investigate the chance of being transferred. Start researching possible openings in other departments. If you feel you have something special to offer your company, perhaps a job could be created for you. Do not overlook that alternative.

You have tried to get yourself out of a present rut and have not succeeded and now have decided that you will leave. It is time for you to sit down with your boss and give your ultimatum: Unless you are given a promotion and a raise, you are leaving. But remember that a promise is not enough unless there is a specific date attached to it, and you must insist on it. What is "soon" for your boss may be too long for you. There are very nice ways of saying you are leaving. You can say that you wish to continue growing but your job does not give you that possibility and your ambitions cannot be fulfilled in your present position. If your boss is unsympathetic, you can still go to personnel to see if you can be moved elsewhere in the company.

Find out how many days' notice the company owes people who get fired; this is an indication of how many days' notice you owe the company when you quit. The higher your position, the more notice you will need to give. The average seems to be approximately two weeks.

It is important for you to leave with a good impression. You never know if this employer will be called for a reference about you. Even if your relationship with your boss was not very good, there are a few things you can do in order to be well thought of after your departure. One is to leave an up-dated job description in order to help whoever will take on your job. You can do this with your boss or by yourself and give it to him or her afterward. You could also offer to stay on until a replacement can be found and then train the replacement, unless you already have a job waiting for you. If you have tried everything to no avail, then do not stagnate. If your situation has not improved by now, chances are slim that it will miraculously improve by itself. It is time to move out in order to move up. It is very often the only way to upward mobility.

It is best to look for work while you still have a job. This gives you more security: You can risk refusing jobs that you feel are not suitable. However, this is not always possible. If you are unemployed and need the money, you can take a part-time job with organizations like Manpower and devote all of your remaining time to the job search. Look into health insurance. There are group plans available, such as the American Association of University Women, or B'nai Brith, and many others. Be wary of employment agencies. They will sometimes send female candidates to organizations that have no intention of hiring a woman, but in order to comply with Affirmative Action, they will say that they have looked and found no qualified women. Head hunters are employment agencies that deal only with executives in a certain salary range. Some work only with people above the $25,000-a-year level; others start with the $60,000-a-year level. I would suggest feminist-owned executive placement services. They seem to do better by women.

Quitting is not a personal failure. Quitting is a show of the following qualities: risk taking, ambition, and willingness to pursue career goals. These are all qualities that would benefit an organization. It is important to point that out when your résumé shows several jobs. Remember also that a second job is easier to find than a first, a third is easier than a second, and so on. The background that you have accumulated makes you that much more marketable.

The second job

You are now in a new position, possibly in a different department, possibly in another company. You're not "entry level" any more, but you are "entering" again. You still have new skills to learn, new people to meet, new norms to figure out. Has anything changed from the first time around? Yes. You will learn everything faster, you have more experience, more confidence, you know your strengths and weaknesses, you are developing a style of your own.

However, do not underestimate the time you need to connect with your new environment. You are not entering as a "learner" but you still need to learn. Your new relationship should be one of *reciprocity*: "I have much to learn, but I have much to offer."

The *Paths to Power* lead us toward more control over our own lives. As you leave the dependent power position and move into an intermediate power position, your tactics change from "manipulation," the generally covert attempts to control or to wield dependent power, to "strategy," a careful plan affording maximum support for your objective. As you obtain recognition and respect from colleagues and superiors, it is possible to look beyond yourself toward the total system. It becomes feasible to strategize to meet your own objectives, but now those objectives are within the larger departmental or organizational goals.

NOTES

1. Jean Baker Miller, *Toward a New Psychology of Women* (Boston: Beacon Press, 1976).
2. Don Murray, "What Makes Readers Read?" photocopied material (Durham, N.H.: University of New Hampshire, 1979).
3. Mary Brown Parlee, "Conversational Politics," *Psychology Today,* May 1979, pp. 48–54.
4. *Ibid.,* pp. 48–56.
5. Gayle Scroggs, "Sex, Status, and Solidarity: Attributions of Inter-personal Touch," doctoral dissertation (Durham, N.H.: University of New Hampshire, 1979).
6. "Exploding a Myth of Executive Job-Hopping," *Business Week,* June 11, 1979, pp. 127–131.

SELECTED READINGS

Bloom, Lynn; Coburn, Karen; Pearlman, Joan. "The New Assertive Woman." *Family Circle,* November 1975.

Carney, Clark G., and MacMahon, Sarah Lynn. *Exploring Contemporary Male/Female Roles.* San Diego: University Associates, 1977.

Ewing, David. *Writing for Results in Business, Government, and the Professions.* New York: John Wiley & Sons, 1974.

Galassi, Myrna Dee, and Galassi, John P. *Assert Yourself—How to Be Your Own Person.* New York: Human Sciences Press, 1977.

Gilbert, Michael A. *How to Win an Argument.* New York: McGraw-Hill, 1979.

Jongeward, Dorothy, and Scott, Dru. *Women as Winners: Transactional Analysis for Personal Growth.* Reading, MA: Addison-Wesley, 1976.

Morley, Eileen. "Women's Thinking and Talking." Unpublished paper. Boston: Harvard Business School Case Clearing House, 1976.

Osborn, Susan M., and Harris, Gloria G. *Assertiveness Training for Women.* Springfield, IL: Charles C. Thomas Publishers, 1975.

Roman, Ken, and Raphaelson, Joel. *The Ogilvy & Mather Guide to Writing Effective Memos, Letters, Reports, Plans and Strategies.* New York: Ogilvy & Mather.

"Sexual Harassment." *Ms. Magazine,* July 1978.

Stone, Janet, and Bachner, Jane. *Speaking Up: A Book for Every Woman Who Wants to Speak Effectively.* New York: McGraw-Hill, 1977.

Tobias, Sheila. *Overcoming Math Anxiety.* New York: Norton, 1978.

Chapter 5

HAND HOLDING

For many years I held
my teachers' hands
and I learned and learned.

Now my hand
is being held
and I teach and teach.

Isn't all this hand holding wonderful?

A Brain to Pick, a Shoulder to Cry On, and a Kick in the Pants

Very few people ever make it alone. We all need someone to lead the way, to show us the ropes, to tell us the norms, to encourage, support, and make it a little easier for us. Who are these people who will do that and where do we find them? They have been called benefactors, godfathers, patrons, rabbis. They are found by men in their "buddy system" or "old boys' network." They can be immediate supervisors or others some steps up the hierarchy. They can be any people with influence inside the organization, or even those outside the organization who have clout.

As you move up from first-line management towards middle management you will need someone who will fight for you in the face of controversy, who can help you bypass the hierarchy, who can open up opportunities for you or see to it that they get opened up, who will help you get inside information or get around bureaucratic red tape. You will need a knowledgeable and powerful friend.

On this part of your path to power, you must have three hands! One to be held by a sponsor as you're being pulled along the organizational ladder; another to pull along those behind you; and the third to hold hands with supportive peers. Let us define our terms and differentiate between them. A *sponsor* is "one who vouches for the suitability of a candidate for admission." A *mentor* is defined as: "a wise and trusted teacher." It comes from the Greek; Mentor was Odysseus' counselor. A sponsor has a *protégée,* "one who's welfare, training or career is promoted by an influential person." A mentor has an *apprentice,* "any beginner, a learner."

The difference between sponsor and mentor is one of function. A mentor will teach you a skill or provide you with the knowledge necessary to perform an identifiable task. Mentoring is focused in the present. A mentor teaches what you need to know now. A mentor may or may not be able to influence your career and need not have any particular clout in the organization.

A sponsor may have very little to teach you about your job, but can help your career by recommending you for special projects, by speaking for you, by taking you along on assignments. A sponsor focuses on your future and must have influence in the organization.

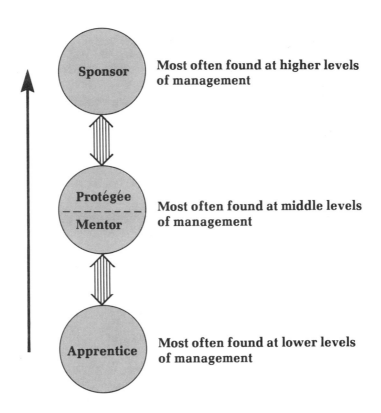

Although mentors and sponsors are at all levels of the organizational ladder, more mentoring needs to be done at the lower levels, while more sponsoring is done higher up. The sponsor's protégée has already proven herself to be management potential, has already arrived at a position of some responsibility and needs another boost to the next level. The relationship between the two is frequently collegial, the protégée often working side by side with her sponsor. A protégée can and should be a mentor to the people below her. A mentor is more of a teacher and helps someone get started, learn the job, know the company norms.

A mentor will help by telling you what issues to focus on at the next regional meeting; a sponsor will help by introducing you to the regional vice president. A mentor sees to it that you gain the competence to become promotable. A sponsor will see to it that you gain visibility and that you are, in fact, promoted.

Sometimes, your mentor and sponsor are the same person; however, it is important to distinguish between the functions to be certain that both are being filled, because *you need both* all along the way. Each time you reach a new level, you need a mentor to teach you the ropes, but as soon as you know them, you need a sponsor to help you reach the next level.

The sponsor-protégée relationship is reciprocal. You are performing better because of this relationship and this will reflect well on your sponsor. To be asked advice or help is ego building for most people, and they, in turn, will be generous with their time and their knowledge.

There are other psychological reasons for becoming a sponsor. Men in their 40s are at a developmental stage where they *need* to sponsor younger colleagues. It is sometimes known as occupational fathering; when they get to their 50s and 60s it becomes occupational grandfathering. The gratification of helping younger people with their aspirations is very great indeed. But there are also power issues involved, for there is real power in backing a promising candidate, in building an effective team. You, the protégée, are also performing a service to your sponsor and to the organization. As you do well and move up, you prove that your sponsor has good judgment.

Mentoring Functions	**Sponsoring Functions**
Present-oriented	Future-oriented
Teaches	Opens doors
Need not have special influence in the organization	Must have influence in the organization
Usually a co-worker or your immediate supervisor	Can be any higher-level administrator
Tells you what you need to know	Tells you what you need to do
Shows you how	Helps you do it
Trusts you with a job	Trusts you with inside information
Gives you assignments	Takes you along on assignments or suggests your name for special projects
Helps by talking *to* you	Helps by talking *about* you
Introduces you to people you will be working with	Introduces you to people who have influence
Teacher/apprentice relationship	Collegial relationship

FINDING A SPONSOR

Although your boss is probably your mentor, he or she is not necessarily influential enough to act as your sponsor. So you will probably have to look elsewhere.

If you want to find a sponsor, *make your goals known,* for people do not automatically assume that women want to move ahead. Women rarely get promotions they have not asked for. *Seek high visibility:* take on projects that will make you be seen; talk about your accomplishments even though people may see you as too forward. It is better than not being noticed at all. Take risks, accept challenges, but be sure you have others' cooperation and approval of your projects. At the same time, gather information about who the people are in the organization who are seen by others as sponsors. You must become a keen observer of which people spend time together: Who usually has lunch with whom? Who seeks and who gives advice? Who hands out interesting work? Who gets the best projects? You will begin to see patterns.

Next time you have a chance to be with a person you have pin-pointed as a possible sponsor, ask some specific questions concerning your work—ask for advice on a project you are involved in or on a decision you must make; ask for some help in a problem-solving situation. If that person has been helpful, be sure to acknowledge it and ask if you can come back other times with similar requests. Do not be afraid to impose a little or to take up too much time. You are showing concern for fulfilling company goals, which is also your sponsor's concern. It is also part of that person's job to help subordinates accomplish results.

It is easier to find a mentor, since it is easier to ask for specific help, than to find a sponsor who will be your advocate. It takes self-confidence to ask someone to speak on your behalf or to ask for advice with your career plans. In other words, you take the direct initiative toward a mentor, but you can only pave the way for a sponsor to take the initiative on your behalf. Be sure to read the signs correctly. A potential sponsor will comment on your work, will appreciate your efforts and be particularly friendly. These are the early signs. You can test them out by following up with a dialogue with this person.

One of the barriers to overcome is the sponsor's probable belief that women are less committed to a career. The sponsor may therefore not want to train a woman as his successor, preferring a man. Women are seen more often as assistants, not as successors. This stereotype needs to be dealt with openly by confirming your seriousness about your career, but it is also important to set boundaries on the demands on your time and energy. Some sponsors may push too hard on their protégées, not always aware of the other role demands women may have. For example, if the woman is married, she may have to deal with her husband's discomfort in accepting her relationship with the sponsor. Sometimes a drink with both sponsor and husband might help demystify the association.

Since men still hold the more influential positions in an organization, your sponsor will probably be a man. What man will risk being seen publicly with a woman and spending time with her in the sponsoring capacity, thus incurring either office gossip or the jealousy of a wife? Obviously, it has to be a man who is *not* threatened by gossip, jealousy, or by the fear of being tempted. It has to be a man who knows that his power comes also from the quality of subordinates, and therefore promoting employees can only enlarge his circle of influence. People will be grateful to him, knowing that he is a person who will support and champion them, and workers under him will perform well knowing that there is a predictable reward.[1]

Very few men below the age of 35 are interested in sponsoring younger people, especially women, but after they reach the age of 40, most men have integrated their masculine and feminine sides well enough to feel free to help a younger female without feeling uneasy about the relationship.

Sexual attraction can be a bothersome issue. You may have feelings about a male sponsor and he may be equally attracted to you. One way to deal with this is to talk about it as a difficulty in your professional relationship. It is perfectly normal to feel attraction; however, not every feeling has to be acted upon. You can both decide that, although the feelings exist, they will not be permitted to get in the way of your mutually enriching professional association. Of course, you may, on occasion, have to attempt a collegial relationship with a boor, who will want to get you into bed. Make it his problem, not yours, and stay away.

The main factor in all dealings with a male sponsor is that you should at all times be totally professional. *Never* be flirtatious or "cute." Even though some men might welcome an opportunity for a dalliance, many would fear the consequences of an office affair. Besides, it always ends up to be detrimental to women.

If you don't fall into the sexual trap, there is still the servant trap to watch out for. Men are used to seeing women serve them, and they may ask you to do all kinds of work that is more fitting for a much younger person or someone in a lower position. It is sometimes difficult to refuse without rejecting. However, there comes a time when you have to sit down with your sponsor and say that you are really ready to move on and do other things. You could perhaps suggest that whatever it is he has been asking you to do can be delegated to someone else.

As more women move into management positions and work side by side with men, the sex objects and the servants will be replaced by friends and colleagues. As men are freed of their stereotypes and consequent prejudices, women will be freed to enjoy the equal recognition of competence and equal power.

WOMEN AS SPONSORS

Women as sponsors are rare to come by: Too few women are in the top positions and influential; many are still trying to simply survive. A woman sponsor has two added qualities to those of a male sponsor. One, she is a role model for you; two, she understands what it is like to be a woman in your position, for she has most certainly been there herself. A woman sponsor can relate to the particular problems of women in organizations and can sympathize with the multitude of roles women must prioritize and play. You may want to find women mentors and sponsors—and indeed, you should have some—but unfortunately, women still have a lower ceiling potential then men. You are best off to hitch your wagon to a star. However, these aren't either/or propositions: Have *more* than one mentor, more than one sponsor.

Queen bees

Older women are frequently known as Queen Bees. What is the definition of a Queen Bee? It is a woman who, in the face of all odds, has made it to the top without the aid of Affirmative Action and equal opportunity laws, of the women's movement, of support groups. On her own, she has become an accepted member of a male peer group. It was probably not just a rocky road for her, but frequently a very lonely one as well.

People who have gone through especially severe organizational initiation rites seem to impose upon new entrants similar, equally difficult experiences. In other words, if they had to earn their stripes the hard way they will see to it that others do the same. To institutionalize a painful experience gives it value simply because not to do so would devalue and diminish that membership. No one wants to place less significance on a particularly painful experience. Perhaps this is what the Queen Bee contends with. We have all often heard the sentiment "I did not have it easy—why should she?" I have heard parents say similar things to their children. The expectation is that the younger and inexperienced people should be mirror images of the Queen Bees, and therefore go through the same experiences.

A good part of this attitude is also derived from the Protestant work ethic so ingrained in most of the people in the United States. The old idea of the connection between the devil and idle hands has, however, been watered down to a less-stringent "hard work is good for you"—it "makes a man out of you."

All organizations have certain rites of passage that every newcomer should learn. However, these need not be the same rites experienced by the women who entered years ago, when it was "so hard." The problems faced by women today are not the same, but this does not mean they are less difficult. There are different expectations today. The Superwoman role is a great burden. Younger women are attempting to do what some of the older women never tried: to be wives *and* mothers *and* professional women. Most of the women who made it to the top echelons of their organizations are single.

However, there is another side to the story. Those women who worked so hard at establishing themselves understandably wish to protect their unique place and the special privileges accorded to the only woman. A younger or newer ambitious woman in the organization can pose a threat. If you are a second or third woman in a place where there is one Queen Bee, show the proper deference to her age, her experience, and her hardships, and maybe this woman can, with time, become your mentor and then your sponsor. But she must see you as a help to her, not as someone who wants to usurp her place.

There are also some women who believe that to get ahead, they need to diminish the accomplishments of other women by making negative remarks about them to men, laughing at sexist jokes or telling some themselves, or ridiculing women who attempt to change male or female sexist behaviors. These women would tend to discount ideas when they come from other women, giving men's ideas priority. They also do not usually confront sexist, racist, or classist behavior in men or women and place men's needs above women's. They're the ones who make statements such as ''I would not vote for a woman president,'' '' . . . go to a woman doctor or lawyer,'' '' . . . trust a woman.'' They express surprise at women's accomplishments and generally react to or relate to women in a condescending manner. Not everyone is educable, so don't waste energy where there is little hope of change.

DIFFERENT TYPES OF SUPPORT

At different times on the path to power, we will need a brain to pick, a shoulder to cry on, and a kick in the pants. We need *a brain to pick* when we need to problem solve, to make difficult decisions, whenever two heads are better than one, or when we wish to share responsibilities.

We need *a shoulder to cry on,* or at least to lean against. It is very hard for many women to break precedents and create new patterns, to become the first female colleagues many men ever had; to be a token; to be stereotyped; and still often, no matter how subtly, to be discriminated against. Sometimes a good talk, a good cry, or someone to listen to complaints is all we need to keep going.

Frequently, women have difficulty asking others to share their burdens. Not that we do not accept the burdens—*that* we do almost too well. But there seems to be a problem differentiating between legitimate help and the spectre of inadequacy. Perhaps we do not ask for help as often as we should for fear that we will be seen as either incompetent or too dependent. Whatever the reason, the too many demands some times become too much for us to cope with, and we begin to withdraw from the fray, to give up or give in. That is when we need to be redirected and challenged. A *kick in the pants* from the right person at the right time and in just the right way will get us going again in our attempt to master our environment, our work, ourselves.

We need a brain to help us *clarify,* a shoulder to *comfort* us, and a kick to *confront* us.* A sponsor or mentor is not necessarily the person who should provide all three. The first will be useful in getting us to think; the second, to feel; and the third, to act. Each one has different capacities and will evaluate us by a different set of criteria. We seldom find all three attributes in a single human being. We are more likely to find one person who is very good at clarifying issues, another who is good at listening and supporting, and a third who knows how to confront you on your weaknesses and will help you develop a plan of action. It is important that you find people who can fill *all three* functions. Do not forget the possibility of spiritual sponsors outside of business. Stop a moment, and see if you can place a name under each. If you can identify specific people, you will know better how to use them when in need.

*Adapted from Janice Eddy, New Dynamics Associates, Laconia, New Hampshire.

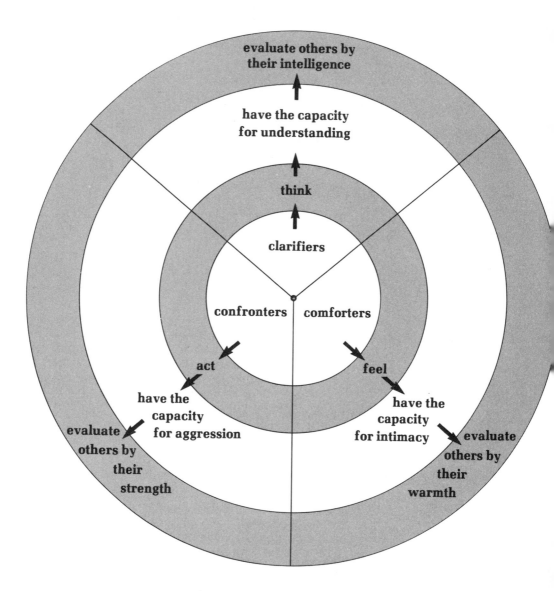

Support groups

Who are they? Where are they? How do we find them? A support group is made up of our allies. It can be our colleagues at work, our friends outside of work, our family, or a professional women's group. It should be as many of these as possible. Ideally there should be both a mixed-gender support group and an all-women support group. Just as men can relax in a very special way when alone with their own sex group, women have a special bonding that occurs when they are together without men.

It may be difficult to break into a support group of your male peers at work. Men, like women, tend to feel more "real" when the opposite sex is not around—even at formal business meetings—so their social conversations may be strained. Many men believe that women are not interested in sports, the stock market, politics, and cars. They feel that the presence of women will cramp their style when they wish to discuss a new secretary's bust measurement, brag about their sexual exploits, or make scatological or sexual jokes. And they are right: Women present *would* cramp their style. Of course, a certain amount of discomfort is to be expected as natural—women would feel the same way if the situation were reversed. However, women must overcome this situation. Why? The critical component is that not only is office gossip shared at such meetings, which occur over lunch, after work, or on weekends, but also a lot of important inside information. This is also where connections are made. It will not be easy to gain entry, but it is critical for women to attempt to get in on some of these collegial get-togethers.

A professional *women's* network is also invaluable. There are "new women's networks" sprouting in many places. Boston has a Women's Lunch Group made up of sixty professional women who meet monthly at the Harvard Club. Women in Business, Inc., in Los Angeles, wishes to help women in middle management or near the top; its criteria for membership is a $25,000 + salary. There is a Women's Forum in Denver, a Forum for Executive Women in Philadelphia (only for managers), and a New Girls' Network of Winston-Salem, North Carolina, which is for *any* interested women. There is also a Business and Professional Women's Club, Inc., in Washington, D.C., for physicians, store owners, etc.

As you see, many of these groups just mentioned attempt to replace the "buddy system" or the "old boys' network." The one in North Carolina is more of a support group. Support groups are very common and their function is different from the "new women's networks." The "new women's networks" provide professional information *and* support, discussing problems encountered at work, providing information about openings, giving references to each other. Simple support groups, on the other hand, provide emotional nurturance and a place where women can talk about problems that have developed with their families as well as with their jobs. The push-pull of both is also discussed. This is not to say that the "new women's networks" do not allow discussion of personal issues. Although the Los Angeles-based Women in Business, Inc. explicitly states that it does not, most others will make room for some discussion around the issues of balancing family and career.[2]

CHANGING SPONSORS

We will need support groups or supportive people all of our lives. For most people, continued growth and development is possible only when their efforts and accomplishments are recognized and respected. In most support groups, the roles are not clearly defined; everyone is available to everyone else, which creates a spirit of mutuality.

Not so for mentoring and sponsoring functions. There will come a time when your relationship with your primary sponsor no longer fulfills your needs—you have arrived at a senior position, or you have a new sponsor who can help in different ways, or the relationship has become oppressive. You may have even surpassed your sponsor. While she or he was unable to let go of the teacher, parent, or guru role, you may be ready for friendship. Some sponsor/protégée relationships can end in bitterness, because the protégée's need for independence and autonomy can make her seem ungrateful. Being aware of the needs your role fills for your mentors and sponsors will make you more sensitive to the best way to part company.

As you move on, you will possess the rare quality of being a senior woman. It is now your turn to be a sponsor. Creating opportunities for younger adults and providing them with support and encouragement may be one of the most significant roles available to people in management positions. Sponsoring makes it possible to transmit knowlege to younger generations. It keeps us in touch with the energy, the doubts, and the excitement of youth. Women in the past have generally been the keepers of tribal traditions. The women of many cultures have always been the story tellers and the socializers of the young. We have always known how to do that well. Today, women have a chance to fulfill those ancient rites again: to admit the uninitiated to the ranks of the adult business world. We have the opportunity to set younger and more inexperienced women on their own *Paths to Power*.

NOTES

1. Rosabeth Moss Kanter, *Men and Women of the Corporation* (New York: Basic Books, 1977).
2. *Business Week*, November 20, 1979.

SELECTED READINGS

Claremont de Castillego, Irene. *Knowing Woman: A Feminine Psychology*. New York: Harper/Colophon, 1973.

Crowley, Joan E.; Leviton, Teresa E.; Quinn, Robert P. "Seven Deadly Half-Truths About Women." *Psychology Today,* March 1978.

Douvan, Elizabeth. "The Role Models in Women's Professional Development." *Psychology of Women* 1–1(Fall 1976).

Kanter, Rosabeth Moss. "Women and the Structure of Organizations." In *Another Voice, Feminine Perspectives on Social Life and Social Sciences,* ed. by Marcia Millman and Rosabeth Moss Kanter. New York: Doubleday, 1975.

Sheehy, Gail. "The Mentor Connection." *New York,* April 1976.

Staines, Graham; Tavris, Carol; Epstein-Jayarahe, Toby. "The Queen Bee Syndrome." *Psychology Today,* January 1974.

Chapter 6

THE SUPERVISOR

I'm in charge of him,
but she's in charge of me.
She is in charge of him
but only through me.

I tell her his needs;
I tell him her wants.
He needs to work less.
She wants him to work more.

And I'm in the middle.

Supervision

Congratulations! You have been promoted and are now a supervisor. You've become a member of first-line management. What does this mean? It means that you are serving two masters, your boss and your workers, and *that* leads to divided loyalties. With whom do you identify? Workers? Management? Undoubtedly, you have come to the job of supervisor through one of three possible channels: Either you came from the ranks, which means you were promoted from within your unit, or you were selected from another department in the same organization, or you were hired for the job from the outside.

The most difficult of all situations is the first—being selected from among your peers. The fact that you have been selected probably means at least one other person has *not* been. No matter how appropriate the choice, no matter how evidently superior you might seem to your boss, your colleagues may not perceive it in the same way.

Most people think that they are at least as good as, if not better than, the people they work with. Certainly most people feel they deserve to be promoted. Therefore, some feelings of jealousy and envy are almost inevitable. I see this occurring among men as well as women. However, for women there is an additional problem. Because of our strong need to *belong,* being singled out and removed from the group to which we belonged gives us a feeling of being apart, of loneliness. You are now the head of your unit and are considered "management." On one hand, you belong to two groups; on the other hand, you belong to neither. You have a different role from the people you supervise yet you do not work closely with the other supervisors on a daily basis. This is the beginning of a new self-concept women must develop in order to survive as supervisors. That concept deals with the issue of authority. You are not a worker any more. You need to give yourself the right to make demands, to check for quality, to reprimand, to help, to teach—to supervise. The advantage of being hired from within your unit is that you know it well; you know its people and they know you. However, they know not only your strengths, but also your foibles. You may encounter problems supervising both close friends and people you have disliked or with whom you have not gotten along well.

If you are promoted from the same company but from a different department, there might be hearsay about you from former co-workers. Some of it may be true, some of it may not be. Another trap you could fall into is the assumption that since you know the company, you will be able to integrate very quickly into the new unit. This is not so. Each unit is different, with different norms, expectations, and objectives. To fit in as quickly as possible, it would be helpful for you to pretend that you came from outside, that you are a new person. This way, the preconceived notions that you may have will not stop you from looking at the people and the job with brand new eyes.

If you have been hired from outside the organization, the disadvantage, of course, is that everyone else knows what needs to be done, how it has traditionally been done, and what norms are followed in the unit. You, however, do not; you must learn them. It is doubly difficult to replace someone else because you are expected to continue *what* the other person was doing in the *same way*, especially if that person was well liked. This is impossible, since you cannot be like anyone but yourself.

There are tactics to help you speed up your learning process. I like to talk one-to-one with the people I supervise as soon as I can. I question each on the norms of the unit, the norms of the company, which are functional, which are dysfunctional—in other words, I ask what the supervision in this unit was like before and what the people would like it to be now. If you supervise a large number of people, a group meeting can be very effective. Even if some of the norms seem dysfunctional at first, such as taking long lunch hours or using the phone for personal matters, I would be very careful not to change them until you have been there long enough to see what the productivity level is, what the quality is, what the working relationships are, and what the level of satisfaction is for most people.

Take your time, look and think before you decide to act on any of the ways things have been done traditionally. Don't forget that you are coming in as a brand new supervisor. You are eager to do well, you want to show your bosses that they were right in putting you in this position, you want to immediately improve your unit, but at what cost? The tendency of some new women supervisors is to forget the people and be very task-oriented in order to disprove the stereotype. After enough time has elapsed for you to understand the workings of your unit, it is important to discuss with your people both the areas that are fine and those that need improvements. This can be done on an individual basis or in a group, depending on the needs that you see. If you make it immediately clear that you want to wait and see how your people work before discussing possible changes, you add to your credibility as a supervisor. Your subordinates will appreciate your consideration in waiting to understand before acting.

SUPERVISORY SKILL

The most difficult part of being a supervisor is that you have been promoted or selected because of your good technical ability as applied to your former job. However, this may or may not evidence an equal ability to manage people. Technical skills are different from managerial skills. In my management training seminars for large companies, I frequently found that some extraordinarily inventive engineers—geniuses in their research and development departments—did not have the slightest notion of how to perform when promoted to management positions. If you were comfortable doing what you were doing, chances are that you would be more comfortable continuing it than instructing, supervising, planning with, and scheduling for others, because these tasks require new skills. There might be a tendency to not let go of what you have done well and to avoid the supervisory requirements in which you have had little or no training and practice.

Is your identity still that of a worker, or are you identified with management? A supervisor identifies with the objectives of management, not with the workers. This does *not* mean that you do not represent their best interest to management—on the contrary, this is one of your roles. *You are the liaison between workers and management*. You push for the workers' needs to be met in all areas—resources, salary, work hours. But at the same time you inform your workers what management needs from them. It is important to be very clear of your identity or you will feel pulled in two directions and become emotionally involved with your workers' problems and with the managerial expectations. Since there can be conflict between what is expected of your workers and what they can in fact produce, your *realistic* and *objective* position is critical to the people and the company.

A Worker	A Supervisor
Works to make a product or perform a service	Sees to it that the product or service measures up to standards and is done within a specific time limit
Has a short time span for making decisions	Must plan in terms of weeks or months
Follows directions given by others	Gives directions to others
Is accountable for only her own performance	Is accountable for her workers' performance
Must perform according to standards set by the supervisor	Sets standards, measures performance, and rewards the workers

SETTING YOUR STYLE

Should your supervisory style fit the job, the people, yourself, or all three? What priority should you give to each element? These are hard decisions. If you face a deadline with workers who do not seem to be as productive as they could be, it might be appropriate for you to be a hard taskmaster. Other times it may be appropriate for you to help someone who is struggling or to give extra time off to someone who needs it.

It is important to note that to be liked *and* respected is very difficult to achieve. Women frequently opt for being *liked*, thinking it the more critical component of the working relationship. It is extremely hard for some women to forego the need of being liked by "everyone"—bosses, peers, and subordinates. In order to be liked, we give praise when it is undeserved, withhold criticism when it would be helpful, and refrain from making necessary demands. This is wrong. In order to accomplish your job, *respect must take a higher priority*. If you are also liked, wonderful. But people will not work well for a boss they like but do not respect. You can earn respect by providing challenge, by treating your workers fairly, by recognizing their efforts and achievements, and by being clear about your expectations.

Who are your workers? Take a look. You will need to behave differently according to the age, sex, ethnic origin, and experiences of all of them. Male subordinates who have never had a female boss may have all kinds of misconceptions about working for a woman. Young people have just finished struggling for independence from their families and may have a difficult time with a woman in authority—she reminds them of their mothers. Older folks may have problems responding to a person in authority who is younger than they are. If your workers are about the same age as you, they may want to impose the norm of democracy ("we are all equal here") and not permit you to exercise your authority.

Women and men have to earn their credibility in different ways. Most people have not seen women in charge, and they must get accustomed to and accept that new image. How are you going to accomplish this? One way is to get your staff together and ask how many of them have never worked for a woman. Then ask those who have never had this experience what their expectations are. What stories have they heard about women bosses? Your openness and willingness to listen will diffuse a lot of the suspicions, misinformation, and possibly even anger. Why do so many men say they will never work for a woman? Think for a moment about the women in most men's lives. They are mothers, daughters, wives, secretaries, maids, waitresses—all serving, caring, supporting roles. You must see how difficult it is for a man who has always seen a woman in a nurturing capacity to suddenly reverse this image. A woman who makes demands, criticizes, has performance expectations, perhaps reminds them of their mothers, and has the power to grant raises and promotions and to hire and fire can be pretty scary for a lot of men.

At the same time, women have a couple of advantages over men that come in handy in the role of supervisor. The first one is that women may be endowed genetically with an ability to perceive small changes.[1] What does that mean? When we see our workers not looking well, not working up to par, or looking distressed, we pick up these messages quicker than our male counterparts and can sooner address the issue. Because we also tend to be better than men at verbal skills, we will also be able to express it in a way that will be heard, as opposed to expressing it in a way that can make the other person defensive or feel dependent.

Some guidelines

I believe you will gain the respect of your workers if you openly discuss these and other matters, be straight, work hard, show your competence, and be fair. Here are a few guidelines for supervisors:

1. This is the most important—notice when something is done well; congratulate people on extra work or care. This way, you will be heard with less negative feelings when you notice work that is *not* done well.

2. When you criticize, be sure to explain in what way the work does not meet the expected standards and provide a plan by which your workers can improve, but don't make it *for* them; problem solve together.

3. Be clear about quality standards and deadlines.

4. Explain the overall objectives in relation to the work performed by your employees. I feel it is important for workers to know how their efforts fit into the overall organizational scheme so that they feel part of a whole, not an insignificant piece lost in some larger unknown purpose.

5. Provide instruction when it is needed.

6. Agree with your workers on some periodic check-in system for measuring progress—one that does not make the workers feel that you don't trust them or are rigidly over-controlling.

7. Give your people a chance to show initiative and creativity whenever feasible, to show autonomy and independence whenever appropriate, but do not give in on quality or performance.

8. Give credit where credit is due.

9. Do *not* talk to one worker about another.

10. Do *not* treat your people as subordinates, but as *your staff*. Delegate as much as possible to show your trust.

11. Do *not* criticize one worker in front of others, or one unit in front of another unit.

12. Do *not* show preference for one or more people.

13. Always support your people. Fight for them and be their champion.

It is possible to create an atmosphere at work that is satisfying for most people. Even if the work has no intrinsic value and is boring, people can gain much satisfaction from colleagueship with co-workers, so allow some time and space for that interaction to occur. If there is nothing positive to say about a place of work, at least people can come to it looking forward to see each other. Although many people at work would prefer to be home, do not underestimate the number of people for whom work is a refuge from a home where there is loneliness or discord. For many people, the work place is the only source of emotional gratification.

FITTING THE WORKERS TO THE JOB

Before you start to fit your workers to the job, you need to be very clear about what your immediate superior expects from your unit. Some bosses are quite vague about their expectations, and women sometimes have difficulty in pushing their bosses to give them concrete information. But you must ask; if you do not, how will you find out? Push on this point relentlessly until you and your unit know exactly what you are supposed to produce, what services you are supposed to render, or what the expected performance is in terms of time and quality. I cannot emphasize this point enough: *If you do not know the expectations, you cannot meet them.*

Your job as supervisor is to maximize your workers' best abilities and to counterbalance the limitations of some by the strengths of others. You need to match the workers with the jobs to be performed. Correct fit here is the critical issue. You do it by analyzing the capabilities of your workers, by planning what needs to be done, and by scheduling who is to do it. These tasks are repeated when you replace several people or reorganize your unit.

Analyzing   From your conversations and observations of the workers in your unit, you should have a good idea of their strengths, their weaknesses, their preferences, and their needs. You may want to jot down some notes on each person so you will have all the information you'll need to plan ahead. What are that person's capabilities? How can he or she help others and be helped in return? What does he or she still need to learn? What is that person capable of doing? What kind of challenge is needed?

Planning   Once you have a good idea of your workers, you can start to fit them to individual jobs. The easiest way to do this is to make a chart, with the names of your workers in the left-hand column and the various tasks that need to be done written across the top.

The final decision as to who does what rests between you and your staff. Remember, you want to keep jobs somewhat flexible to allow your staff room for challenge, creativity, success, and, if necessary, improvement. If there are jobs that no one wants to do, these could be rotated among everyone, keeping in mind the prerogatives of seniority. The same holds for the preferred jobs, if there are several equally able people vying for them. You may also have a person who is not yet able to do a certain job, but wishes to be trained for it in order to advance. All of this information must be elicited from your workers, and all you need to do is ask and then listen. Not everyone hopes to move up, but those who do are grateful for a boss who is interested and wants to help with those career objectives. Be sure, however, that you both agree that the objectives are realistic.

Scheduling   You now have a list of people performing a list of jobs according to their competencies and desires. Let us complicate matters. Suppose each one of your workers is responsible for many duties, or has a job with many components. You should make a list of all the activities related to each job. This would look something like the checklist on the next page.

Using such a chart, you can check to make sure workers are doing what they are supposed to, that they are not doing someone else's work or skipping a task they perceive as someone else's, and that the sequencing of their activities is logical. You will also know at a glance if gaps exist. By including a time frame for each activity, you can also judge whether they are taking too long or too short a time doing the tasks. What you have just done is called planning and scheduling. You have decided what should be done, you have identified who is to do it, and in what period of time.

Another important way of scheduling is to look at (a) what needs to be done on a *daily* basis, such as reading the mail and returning phone calls; (b) what needs to be done on a regular *recurrent* basis (every second day, once a week, or once a month), such as writing monthly reports and taking inventory; and (c) what needs to be done *uniquely* (only once in a while), such as interviewing a new applicant. Be sure that the work load is equally divided so that the same people don't always do the usual daily things while a few others always do the unique or recurrent things. It is interesting to note here that if you study the relationship between job category and gender, you will usually find that women do the daily tasks and men do the recurrent and unique duties. So watch that your people do not get stuck in the same discriminatory rut. Have your workers prepare activities charts, along the line of this sample, dividing them into daily, recurrent, and unique duties.

Activities Chart

Activity	Daily	Recurrent	Unique

Checklist of resources in a doctor's office*

Jobs:	Answers phone	Schedules appointments	Types bills, forms, and correspondence	Files	Updates patient records	Assists doctor	Cleans exam room	Performs minor lab procedures	Examines patients
Employee:									
Receptionist	X	X							
Typist			X	X					
Working Supervisor					X	X			
Nurse A						X	X	X	
Nurse B						X	X	X	
Doctor A									X
Doctor B									X
Doctor C									X

Gaps:

1. When the receptionist is out sick, who will answer the phone and make appointments?
2. If the typist is on vacation for a week and the working supervisor is ill, who will type up the Blue Cross forms that need to go out weekly?
3. The filing has been piling up because the typist has been so busy. Who can help file?

*I would like to thank Maruta Mulins for developing this chart.

Project planning

I would like to now explain the Gantt Chart, a very useful instrument for project planning. It can be put to use when you move your office, build a home, or relocate people or equipment. The chart shows the days or weeks in question and fits in what needs to be accomplished in the allotted time. To fill in the chart, mark in first what needs to be done last, then move backwards along your list until you get to the beginning. For example, you cannot paint a wall in the house you are building until the plastering is done; and you cannot plaster until the plumbing and electrical wiring have been put in; and you cannot do that until you know where you need the outlets; and you cannot make that decision until you know the position of the furniture; and you cannot know that until you determine the purpose of the room. It is a good idea to also write in the name of the person in charge of each activity. I like to see that chart enlarged and posted on a wall for all to see, especially for a project involving many people.

HIRING

If a new job is created in your unit, you can either *hire* a new person to do it or *train* someone already on your staff. If you hire someone, you should have an idea about what the job entails and what specifications it requires. What information do you need from the applicant? You only know this if you have made up a list of the qualifications necessary to fill the job. Remember the chapter on being interviewed for a job? This is the reverse side of the coin.

Job appplicants are commonly nervous, so do something to ease the tension. Chat about how they heard of the job, or how they got to your company, or the weather. Offer a cup of tea or coffee. If I see a very tense person in front of me, I say things like, "I'm sure you must be very nervous; interviewing is not easy." This may help the person to relax. I like to ask open-ended questions, which cannot be answered with a "Yes" or a "No" but allow the person to talk. "Tell me about yourself and your work experience," "Tell me what you like to do and what you don't like to do," and "What do you expect from this job?" are good examples. You must provide enough information to give the person a clear idea about the job, your expectations, the co-workers, and the work atmosphere. It's better to forewarn people at this time of any potential difficulties on the job than to surprise them once they're in, which can lead to anger or disappointment. If candidates look promising, I take them around, introduce them to a few people, and generally make them feel welcome. Do not underestimate the importance of co-workers. A study of young white-collar and blue-collar workers found that the first item on the list of what they like to have at work is friendly co-workers.[2]

It is important to know which questions you *can* legally ask, and which are illegal (see Chapter 2). For example, I tell the applicants, "I do not ask any personal questions because it is illegal; however, please feel free to tell me anything you wish me to know." This lets the applicants say whatever they feel might be appropriate. When-

INTERVIEWING GUIDE*

Ask yourself about:	Ask the candidate
Initiative	
A self-starter? Completes own tasks? Follows through on assigned tasks? Works independently?	How did you get into this line of work? Do you prefer to work alone or with others? What do you like and dislike about your kind of work? When have you felt like giving up on a task? Tell me about it.
Motivation	
Settled in choice of work? Uses some leisure for self-improvement? Willing to work for goals in face of opposition?	What ambitions does your spouse (or others) have for you? What have you done on your own to prepare for a better job? How will this job help you get what you want? What obstacles are most likely to trip you up?
Stability	
Excitable or even-tempered? Impatient or understanding? Uses words that show strong feelings? Poised or impulsive? Controlled or erratic?	What things disturb you most? How do you get along with people you dislike? What were your most unpleasant work experiences? most pleasant work experiences?
Attitude	
Competes without irritation? Bounces back easily? What is important to him or her? Loyal? Takes pride in doing a good job? A cooperative team player?	Did you ever lose a competition? Tell me about it. Do you feel you've made a success of life to date? How? Who was your best boss? Describe the person.

INTERVIEWING GUIDE*

Ask yourself	Ask the candidate
Planning	
Able to plan and follow through? Depends on supervisor for planning? Able to fit into company methods? Able to think of ways to improve methods? Sees the whole job or gets caught up in details?	What part of your work do you like best? least? What part is the most difficult for you? Where do you want to be five years from now? If you were the manager, how would you run your present job?
Insight	
Realistic self-appraiser? Desires self-improvement? Interested in others' reaction to self? Takes constructive action on weaknesses? Takes criticism well?	Tell me about your strengths and weaknesses. Are your weaknesses severe enough to correct? Why or why not? What is the most useful criticism you ever received? From whom?
Social Skills	
A leader or follower? Interested in new ways of interaction? Gets along with most people? Will wear well over the long term?	What do you like to do in your spare time? Have you ever organized a group? Tell me about it. What methods have you found to be effective in dealing with people? What methods are ineffective? What kind of people do you get along with best?

*Adapted by permission of Management Recruiters International, Cleveland, Ohio.

ever feasible I also like to tell the applicants *not* hired some of the reasons why I did not hire them. Besides being common courtesy, it will help these people do better in the next job interview. However, be careful to relate your comments only to the needs of the job or you may be faced with charges of discriminatory hiring practices.

There are many questions that you ought to be asking yourself about a prospective employee. Some of these, such as which technical skills the person has, may be easy to ascertain. Others, such as whether the person has leadership ability or willingness to learn, may be more difficult to determine. The guide on pages 114 and 115 can help you ask the right questions. Now listen carefully to the answers.

TRAINING

The first day your new employee is on the job, be sure that someone is responsible for the person if it is not yourself. Someone should show the new person around, introduce the co-workers, and relate the basic necessities: where the washroom is; where, when, and with whom breaks and lunch are taken; transportation concerns, including parking and car pools; special office procedures, such as how to operate certain equipment; whatever you feel might be helpful. A way to get in touch with what needs to be told is to remember your own first day on the job. How was it for you? What did you wish you knew?

Now comes the training period. The new employee needs not only to be *told* what the responsibilities of the job are, but also to be *shown*. People learn much faster when they are shown something than when they are told about it, and they retain the knowledge much better if they can *practice* what they have been shown. There are two important things, to remember: One, the easiest type of job for you may be very difficult for someone else. Two, many people have never dealt with a woman trainer or manager. If you sense any reservation, ask if this is the first time that person has worked for a woman. If the answer is no, fine. If the answer is yes, you have already helped to bridge the gap between you by showing that you are aware of the lack of role models, and that you are open and willing to talk about it and not be prejudiced. Whether you are a younger woman, an older woman, or a particularly attractive woman, you will probably elicit some kind of feeling. Be aware of the problems your gender may cause for others, especially traditionally raised men. Accept their difficulty as a result of the socialization process. Defensiveness or anger on your part will only engender defensiveness on theirs. Instead, be patient, tolerant, and open.

I have a firm belief that people who act unkindly, angrily, or insensitively do so not because they wish to, but because something is hurting, or something is scary, or they feel doubtful or insecure—they are are not feeling good about themselves. When I worked as a therapist, I learned that there are no unnatural mothers or bad spouses out of choice. The people who do not know how to act better are damaged or wounded. They act out of defensiveness, to protect themselves, believing that the world is hostile and they had better strike first or they will be victimized.

DELEGATION AND MOTIVATION

Well, you are all set. Each employee knows what to do and is aware of your standards: what is good, what is acceptable, and what is poor. Your time expectations and deadlines for the various tasks are just as clear. Now it's time for you to make a list of your own activities and start looking for tasks and responsibilities you can delegate to others. You too have your special abilities and special limitations. It is important for your strengths to be used well and equally important for you to get help with what you do less well.

Many women have difficulty delegating tasks for several reasons. First of all, women have traditionally been delegated to. We "do" for our parents, for our husbands, for our children. We've been taking care of everybody! Second, women have always had the emotional responsibility of most jobs in the traditional place of female power—the home. This is a difficult habit to break. Do you supervise your children closely, or do you trust them to do their chores well enough? The kind of supervisor you are at home is probably the same kind you will be at work.

Third, many women think of work as merely a job as opposed to a stepping stone in a career. We get obsessed with the job to be accomplished and do not see it as one of many jobs in a career continuum. We should attempt to delegate as many facets of our job as possible so that we can go on to more interesting things, such as rethinking procedures and staffing needs and being generally more innovative and creative. We know from research that the more creative a manager, the more successful she or he will be.* You must first act like a manager before you can be promoted to one.

So how do you delegate? Take a look at your people and see if anyone (1) is ready to take the job on right away, (2) might be ready with some additional training, (3) will need a lot of additional training to do it. It *is* worth your while to train people to take on some of your responsibilities. If you are the type who says, "I don't have time to train anyone because I have so much work to do myself," this is a clue that you *really* need to train others. The easiest way to begin is with what you do on a fairly routine basis.

Delegation is important in motivating your staff. It creates new challenges for them and shows that you trust them with additional responsibilities. Every time you let one of your employees make a decision, initiate an idea, carry something out on her own, you take a risk. If you are afraid to take that risk, then weigh your resistance in terms of what is the worst that could happen if your employee failed. Sometimes the worst is not that terrible, and therefore worth the risk. Allowing them the freedom to take risks at work is often highly motivating for employees.

How else do you motivate your workers? You ask the right questions. What are their needs in terms of recognition, esteem, friendship, autonomy, self-realization? Where are they in their lives? Do they prefer the security of being told what to do or the challenge of participating in the decisions? Does each employee fit the job, and does your style of management fit both the person and the job?

---

*David Ley, unpublished doctoral dissertation, Michigan State University, 1978.

Don't sit in your office and wonder how to get the answers to these questions—meet with your people and ask. You get to know them by observing them. Because when you come down to it, motivation is really very simple: What rewards do your people need in order to attain the objectives of your organization? These can be in terms of raises and promotions, which are what we call the extrinsic rewards. They can be intrinsic, rewards from the learning, growth, and satisfaction people feel from a job well done, or the good feeling derived from being part of a unit whose objectives they believe in and whose successes they can share. Don't you work better when you get an approving nod or when you understand the full significance of what you are doing? Ideally, there should be both, the challenge of high standards to meet and success at meeting them.

Learning from success

If you find yourself doing a lot of criticizing and not much approving, something is wrong. *People learn more from their successes than from their failures!* I'll say this again: PEOPLE LEARN MORE FROM THEIR *SUCCESSES* THAN FROM THEIR FAILURES! Remember this fact. Notice good performance as well as poor performance. It is amazing how many supervisors don't do that. A study of whether satisfaction causes performance or performance causes satisfaction found that good performance determined rewards, which in turn produced satisfaction.[3] The research indicated that the *only* link between satisfaction and performance was through the reward system. This is an important concept for managers to remember. It means that to get good performance you must give a reward. This can take the form of praise; it does not always need to be material. Reward based on current performance seems to positively affect subsequent performance. The problem, of course, is the appropriateness of the reward, because what one person sees as a reward may not be seen the same by another.

In order to motivate your staff, you also need to understand their problems. What are they? The job could be monotonous; there may be no opportunity to try out new things, no way to experiment with new skills; the people may feel incompetent, unrecognized, not trusted.

These feelings get expressed in two major ways. The first one I call *disengagement*, which is manifested by depression. A disengaged employee is apathetic; does the minimum amount of work, just going through the motions; feels incompetent and therefore never initiates anything; and feels generally undervalued and is therefore an oppressive presence in the department. The other way of reacting is through *anger*. This can be a direct expression, such as the case of the fuming person who throws things around, slams doors, and barks at people. It can also be expressed indirectly, which is called passive/aggressive because the aggression is covert. This is manifested by often repeating the same fault; by ''misunderstanding'' directions; by doing things to the letter, which means not doing them well; by complaining behind your back and inciting other workers to rebel.

How do you deal with these two categories? Disengaged people need to be *re*-engaged, excited, and challenged; they need to believe in something, and they also need to believe in themselves. Angry people need to express their anger, to identify its source, to be helped to legitimize its expression. Both categories need to problem solve with you as to what can make them more motivated or less angry. The problem may lie in the job itself, in the relationship that they have with their co-workers, in your own supervisory style, in the policies of the company, or in something that is happening to them outside work. You cannot help if you don't know. It sounds very simple, but very few people actually sit down with the worker and say something like, "I see that you are upset, or angry, or depressed. I'm wondering if we can talk about it." Once you've done this you have to *listen*—listen to what they have to say, to their suggestions. And remember—if you use their suggestions, give them credit; this, too, is a reward. If you do not, explain why.

My assumption is that most people want some measure of control over their lives; so if they participate in the decisions of how to meet their needs at work, they will be more committed to the results of these decisions. Do not promise what is not feasible, but explain why you feel it is not feasible. You may wind up agreeing only to disagree, but at least the area of disagreement has been identified and discussed, which will have less of a chance to lead to disengagement or anger. Sometimes things are neither in their control nor in yours. Having your sympathetic ear and knowing that you understand their predicament will be helpful.

PERFORMANCE EVALUATION

The purpose of a performance evaluation is to give you a format for an ongoing program of communication. Most companies have formal procedures for you to follow or to use as guidelines. If you do it automatically, just to meet the requirements of your organization—that is, you talk with each employee once a year, check the items on a form, have the employee sign it, and send it in—you are not doing your job effectively. Its purpose is not to comply with company regulations once a year; it should be used as an incentive to keep the communication active *throughout* the year. Although I am not talking specifically about management by objectives (MBO), there is a small part of MBO that fits nicely into the purposes of a performance evaluation.

The overall theme of a performance evaluation is to identify the components of the job and how your workers can meet them. Discuss how well the employees met them in the past and what needs to be done to meet them as well or better in the future. Which of *your workers'* objectives can you agree with, which of *your* objectives for your workers can they agree with? This is critical. You must agree upon objectives. An important component to identify is what you as a supervisor can do to help. But it is not enough to look at which objectives have been met successfully and which ones need more work. You need to look at objectives in a time frame: When can you both expect to have the work completed? How will you know if standards are met—how will you measure or evaluate? Will progress reports help your employee? Most jobs can be divided into smaller components, which are easier to handle than the total project as a whole. Help your workers do this, but don't do it for them. Ask them to divide the work into feasible segments on their own and to show you what they have done. Give them as much responsibility as you can in figuring out their own solutions. You may need to help with suggestions, but your job is to pose the questions and then be available to evaluate *their* solutions.

The first step in a performance evaluation is done before the employee is called into your office. Sit by yourself for whatever time you need and think about your employee. Do you like the person? If yes, what do you like about him or her? If you dislike the person, why? Very often, we react to people not because of who they are or even how they are doing at work, but because these people remind us of others. I know my own danger signals: If I am anxious about doing a performance evaluation, if I feel some fairly strong emotion, then a little flag is raised inside of me that says, "What are you reacting to?" I pay attention to this and really try to think through what is going on inside of me. It's very important to not be controlled by my emotions, so that my responses to that person are fair. The other danger signal I recognize in myself is when I think about the person in terms of what she or he *is* as opposed to what she or he *does* on the job. The performance evaluation is only about the *performance*. It does not evaluate the person. I ask myself the questions, "What do I need to talk about with this person that might be helpful? What are my objectives for this interview? How would I like this person to feel when he or she walks out of here? How do I accomplish that?" As I think, I make some notes for myself. Only then do I feel prepared for the interview.

A two-way street

A performance evaluation is a two-way street. Not only should you tell your subordinates how you feel about their work, but also ask them in what way you can be more helpful. Some of the questions I asked my secretary during her last performance evaluation were whether she can read my writing, whether I give her enough lead time, whether I am ever too demanding or too much in a rush for things, and whether she ever feels that I have inconsiderately overloaded her with work. In other words, I want to know if she thinks I am a good boss.

However, this is not how to start a session. Employees are anxious about the interview and want to know as quickly as possible your perception of *their* work. But before you share this with them, ask them what *they* think of their performance: Are they generally satisfied or dissatisfied with their productivity? What do they think of the job in terms of variety, challenge, and security? How do they get along with their co-workers? At some point in the interview discuss their career paths and what they need to do to get where they want to go. Then talk about your perception of their performance. Always confirm your observations, both positive and negative, with data.

For a man, remember that you are doing something that may well be unusual in that man's experience: A woman is evaluating him. That probably hasn't happened since he was a little boy, when his mother and his grade school teachers told him what to do. It is important for women to know that the resistance they may get from men may simply be a reaction to their femaleness in this particular situation. Again, the best way to handle this is to *talk about it.*

I would like to share with you a way of handling problems that has worked well for me. I try to understand the problem, attempting to diagnose it in my head. If I have a hunch about the cause I disclose this hunch to the other person. I test it, saying it is only a hunch. I risk being wrong, but that's okay, I'm not perfect. We can then discuss it. So often most of us *know* the answer inside, but we don't come out and say it, we don't risk testing the thought or the feeling. I urge you to try it. Do it first in a safe situation. As you get more skilled, it will serve you well.

It is not pleasant to criticize subordinates, but it is especially difficult for women because we risk incurring dislike. A well-thought-out performance review should help your employees understand not only your expectations but also your view of their strengths and weaknesses in terms of both the job to be done and their future career plans. If you can praise and criticize constructively, you will earn the *respect* of your people.

THE PROBLEM EMPLOYEE

Every supervisor at one time or another must face problem employees—people who consistently make mistakes, who are always late, frequently absent, slow, or generally difficult to get along with. There may be family issues or a drinking problem; perhaps the person has poor interpersonal skills or a limited intelligence; maybe he or she is carrying a chip on the shoulder. What can you do? There are three ways to handle it. The first is the *preventive approach*. If you have a problem employee, go back to your initial selection and screening process and review your interview notes. Was there any indication of potential problems? Early detection is the best prevention; talk with the employee at the first signs of an impending problem. The *therapeutic approach* means counseling the person, offering more or better training, or getting outside consultation for the employee or for yourself. The *punitive approach* is harsher: You demote the person, lower the pay, withhold raises, transfer the employee to another department, or fire him or her.

Use the punitive approach only after you have tried the therapeutic approach without success. If you have several problem employees, you should pay serious attention to how you staff and implement the preventive approach. Most women find it hardest to use the punitive approach and easiest to use the therapeutic approach. However, it is very important to know when keeping a problem employee on the payroll is no longer functional; to know when there is relatively little chance of helping no matter how much time and energy you expend. The problem is deciding between human and organizational needs. You must take both into consideration, and advice from your heart and your head may not always coincide. If you are troubled and not sure what to do, talk it over with your colleagues or your boss. Use others to help you problem solve. Get support for your decision and you will not bear the consequences alone.

When you talk to a problem employee, base your conversation on observed behavior, not on hunches or feelings. You must always be able to back anything you say with actual examples. For instance, instead of saying, "I don't like your attitude," present an example, and a time when it happened: "When I try to explain something to you, you frequently interrupt me as you did yesterday, by telling me that you know this already.'' This helps the person know exactly what you dislike. If you add, ''And when you do this, I feel that you don't want to listen to what I have to say, which bothers me,'' you add another dimension: You explain the impact that person has on you. Because I think this is so important, let me give you another example. *WRONG WAY:* ''I don't think you are very committed to this job.'' *RIGHT WAY:* ''When you start looking at your watch impatiently half an hour before closing, and then dash out of here like the place is on fire, I think you don't like working here and I feel badly about it.'' You could add, ''Perhaps you have a good reason to be in a hurry, and I am unjustified in attributing your behavior to something that is going on here, but I would really like to discuss this with you. Is now a good time?''

Make sure that the problem employee knows exactly how the performance is to be improved and set a time limit. It's a good idea to write it all down; give the employee a copy, and keep one for yourself.

Firing, a last resort

You have used the therapeutic approach unsuccessfully, you have used the punitive approach to no avail, and your problem employee is still a problem. As difficult as it is, sometimes the only solution is to fire a nonproductive employee or one who lowers the general morale, or creates more dissention than is comfortable. This is a difficult decision to make, especially if the employee needs the job or has a bad situation at home. But keeping someone on the staff out of pity rarely works for the employee or for the organization.

Once your decision is made, be firm. Take the time to listen to your employee's anger or hurt, and explain the reasons for your decision. You may never really be heard accurately, but do it anyway. Do not allow any ambivalences to be detected.

When the problem is caused by a poor fit between worker and job, it is helpful to talk about your employee's career directions. Be fair and objective; being vindictive may offer you some relief from frustration, but it does not help your reputation as an understanding supervisor. You can work toward the company's goals and still look out for your employees' interests, giving your people more than one chance and trying to correct bad match-ups between worker and job. This, after all, does not mean that the employee is a bad person, it means that he or she is not *suited* for that particular work, in the particular climate of your unit, under your particular supervisory style. The fault may be in many places. Be attentive to that possibility and remain supportive to the end. Yes, you can even *fire* in a supportive manner! Your style makes a lot of difference: A discharged employee could feel really destroyed or feel that elsewhere he or she may have a chance to do better. However, there is no way, really, to terminate someone without incurring hurt.

If firing is the only solution, check the company's firing policy before you do anything. If your company has a union, your handbook will list proper procedures in detail. These procedures must be followed *to the letter*—that is, you may be required to give three written warnings, with copies to management and union. Be aware of any requirements before you act. Give the person time to look for a new job—*unless* he or she is so destructive to the work climate that letting them go on the spot is called for. If the employee requests a reference from you, try my system. I never write poor references. If they can't be positive, I refuse to write them at all.

As a supervisor, you are in charge of your employees and are responsible for their output. If you do not pay attention to their level of satisfaction, in terms of both physiological and psychological well-being, you are not doing your job. This does not mean that you can't set high expectations for productivity. On the contrary, you *can* be a strict taskmaster as well as a caring person, you *can* have high standards and an understanding heart. What kind of supervisor would *you* like to have? Well, be that supervisor for others.

NOTES

1. Eleanor E. Maccoby and Carol N. Jacklin, *The Psychology of Sex Differences* (Palo Alto, CA: Stanford University Press, 1974).
2. Daniel Yankelovitch, *Changing Youth Values in the Seventies: A Study of American Youth* (New York: McGraw-Hill, 1973).
3. Charles N. Greene, "A Causal Interpretation of Relationship Among Pay, Performance, and Satisfaction," paper presented at the Annual Meeting of the Midwest Psychological Association, Cleveland, May 1972.

SELECTED READINGS

Diamond, Helen. "Wanted: More Women in Management." *Educational Horizons* 53-3(Spring 1975):125–128.

Gordon, Raymond L. *Interviewing Strategy, Techniques and Tactics.* Homewood, IL: Dorsey Press, 1969.

Kay, M. Jane. "A Positive Approach to Women in Management." *Personnel Journal,* January 1972.

Richardson, Stephen A.; Dobrenwend, Barbara Snell; Klein, David. *Interviewing: Its Forms and Functions.* New York: Basic Books, 1965.

Steers, Richard M., and Porter, Lyman W. *Motivation and Work Behavior.* New York: McGraw-Hill, 1975.

Steinmetz, Lawrence L., and Todd, H. Ralph, Jr. *First Line Management.* Dallas, TX: Business Publications, Inc., 1975.

Van Dersal, William R. *The Successful Supervisor* (rev. ed.). New York: Harper & Row, 1968.

Vogel, Alfred. "Why Don't Employees Speak Up?" *Personnel Administration,* May–June 1967.

Chapter 7

DUAL-CAREER FAMILY

When he brings home the bacon
She fries it.

When she brings home the bacon, too,
They eat out.

The Balancing Act: The Multicommitted Woman

Singly committed men and women whose priorities are clearly with work outside the home, and singly committed women (and a few men) whose priorities are clearly with homemaking, do not experience the push-pull that multicommitted women and men must face almost daily. "Multicommitment" refers to the responsibilities undertaken that create more demands on time and energy than is normally available. *The more equally distributed the priorities between the commitments, the more conflict and stress is produced.* Women and men who balance family and career, who juggle earning a degree at night with work and/or family, whose energies must be spread over several equally important endeavors, are in the bind of constantly having to choose one over another. Criteria for choice are seldom clear cut and what does get done leaves something else undone, or else doing one thing well usually means doing another thing poorly. If the attempt is made to do everything equally well (the Superwoman or Superman), the price for it is paid in terms of exhaustion and no time for leisure, for relationships, or for oneself.

All of us make constant choices in our daily lives, but within a specific frame of reference that dictates priorities. For most people (usually men) who value work above family, a sick child will not be an issue. For most people (usually women) who place an equal value on work and family, the sick child will present a basic unresolvable dilemma. Getting someone else to care for the child might generate guilt (bad mother); not going to work might produce a serious handicap to career aspirations (not dependable). Either way, true multicommitments produce stress. In our society, the more and more women entering the work force are the majority of the multicommitted. How to balance the traditional commitments to family and the new commitments to a career become critical issues for women who are facing these roadblocks of responsibility and expectation on their *Paths to Power*.

Adults go through different periods in their lives, called life stages, life phases, life cycles, or developmental periods. To some extent, these periods are predictable according to chronological age. However, for a woman, age is not the only major dimension in her life cycle; her *family situation* is another important factor. Whether she is married or single and has one, two, or more children influences her much more critically than her male counterpart, as do the ages of the children and the spacing between them. Most husbands tend to pursue their careers independent of their family situations, except perhaps for taking fewer risks when there is a family to feed. Whether their wives have children at age 20 or 35 will make relatively little difference to their careers. For women, it will make *all* the difference.

If a woman of 35 has a 3-year-old child, a baby at home, and a full-time job, her issues are very different from what she would face if she were 35 with a 15-year-old, or 35 and single with no children.

In other words, her family situation will dictate her options much more than has ever been the case for men. Based on the distribution of her commitments, whether she decides to remain single or marry, to remain childless or interrupt her career for children, to delay her child-bearing for a career or integrate both, her issues and concerns will be totally different. There are three general categories based on economics:

1. A woman may share two incomes with her partner; they are dependent upon each other's earnings. She may be childless or have children.

2. A woman may live on only her own wages; she is totally independent. She may be single and childless or a working mother.

3. A woman may pursue a traditional homemaker's role; she may be totally dependent upon her partner's income or only partially dependent if she holds a part-time job.

The first two categories show the multicommitment; the last is usually singly committed unless she is engaged in volunteer activities or works part-time.

This chapter will examine the issues facing women in those different life situations. Age is a factor in three circumstances. The first is the biological ceiling imposed by nature on childbearing (although women are having children later than ever before). The second is society's stereotyping reactions to age (for example, the highest rate of unemployment is among middle-aged women).[1] The third is when women are told they are ''out of synch''* with traditional career patterns as far as age is concerned (the assumption being that the ''normal'' is the uninterrupted male career path).

Just as roles, tasks, and resources influence behavior at work, roles (work and family), tasks (responsibilities), and resources (income and support) influence personal expectations and responses (see Appendix D, pp. 262–269).

SHARING TWO INCOMES

The dual-career family, the two-career couple, the double paycheck—all these refer to the increasing phenomenon of married women entering the labor force. Even

---

*A term to indicate that the person is at a different personal and/or professional stage of life from what is ''normal.''

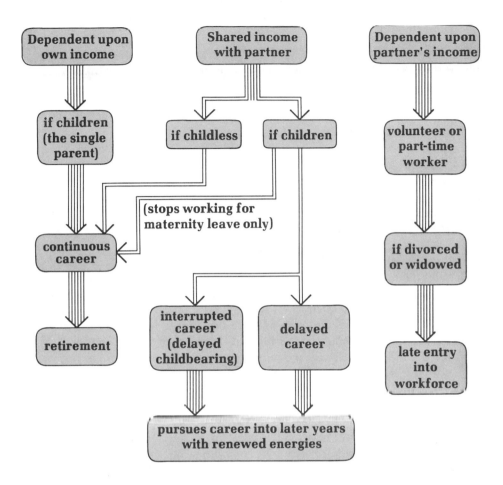

though income is now earned by both partners, the issues around commitment are frequently still in a stage of transition between current attitudes and traditional values. What we often find in this type of couple is a singly committed husband (commitment is to a career) and a doubly committed wife (commitment is to *both* marriage and career). As a matter of fact, if there are children, the woman may even be triply committed: Her roles are worker, wife, and mother.

If we include a commitment to continuing education (she may be going to evening classes or working part-time and earning a degree), we add still another role, that of student. Now we have a quadruply committed woman. I suggest we speak of today's couples as multicommitted couples. The husband may also have three roles—husband, father, worker—all with equally high priorities. A lot of energy and time are given to these various roles, which are frequently in conflict. A woman's mother, her child, her husband, her boss, and her teacher, can each expect total commitment.

Career versus job

Half of all husband-wife families have both spouses working, and almost half of those working wives are working full-time, year-round jobs. The wife generally contributes about 40 percent of the family's income, but even when her salary is the same as or exceeds her husband's, she is still seen as the *secondary* wage earner. This attitude centers around the perceived difference between a career and a job.* A career has educational prerequisites for a profession or includes training for a particular expertise. A career has a professional goal. The primary purpose of a job is to earn income. It has limited advancement possibilities, permitting little or no further growth and development for the individual. A job can become part of a career path, but this is to a large degree a matter of aspiration.

The importance of the distinction lies in the way a woman perceives herself and is perceived by others.† If *he* is on a career path or is studying for a career and *she* has a job to either supplement the family income or to support him while he's studying, *they* will tend to set priorities in terms of *his* needs. She will take on the female roles: She handles the many social obligations (keeping a calendar, making dates, writing invitations and thank-you notes, making phone calls around social occasions) and the entertaining; if there are children, she is the obvious child-care parent—she will even miss work if needed; she is also seen as more responsible for home maintenance. Sometimes, taking care of the relationship with both sets of parents is added to her responsibilities. She writes letters and visits—doing *for* and doing *with* as necessary.

Sharing the housework

If she sees herself as having a career, being challenged, having potential but he does *not* see it that way, they will have trouble in terms of mutual role expectations. She will expect him to share home-related tasks; he will not. But even if he *does* see her as being on a career path, too, the man still tends to see himself as the primary wage earner, no matter what the actual salaries may be. This still produces some problems around who does what, when, and how. A multicommitted woman needs the moral support of her husband to deal with the many responsibilities she assumes. But unless he believes in the value of her commitments, she will end up with more than her share of the work.

A woman in the dual roles of wife and worker needs to ask how committed her husband is to sharing household chores. If they have the same work schedules, who makes dinner? Should they do it together or take turns? Should one cook and the other clean? The time to discuss values, priorities, and expectations is *before* marriage, but the discussion must continue throughout the marriage.

There are many ways of dealing with the division of labor. The important point is to make sure that neither person feels entitled to do less work in the house. By talk-

*Douglas Hall defines "career" as "the individually perceived sequence of attitudes and behaviors associated with work-related experiences and activities over the span of the person's life." (*Careers in Organizations,* Goodyear, 1976).

†Israel passed its Equal Rights Amendment Law in 1951. In 1953 there was a change in the Marriage and Divorce Law: The word "bacal" (husband, but also owner or master) was replaced with "ish" (man), thus eliminating any implication that women are inferior.

ing often with each other, most couples should be able to come to a mutually agreeable arrangement. Even though many couples would agree that housework is boring, *when shared it becomes less alienating.*

It matters little who does what. What matters is that each feels the other is doing a fair share and that neither feels unfairly overloaded. I know of only one way to accomplish this—to keep sharing your feelings and keep renegotiating the various responsibilities. Working on a marriage or a partnership takes teamwork—it takes *dual effort.*

Practical versus emotional responsibility   In the negotiation with your husband as to who is in charge of what, the words "in charge of" are all-important. Husbands tend to have a "help out" attitude, which shows two very different senses of responsibility: practical and emotional.

There is a widely held assumption that, since convenience appliances are readily available, housework now takes less time. The assumption is wrong: Women in the United States spend more hours cleaning today than ever before, probably because of our higher standard of living, which requires higher standards.[2] The new food processors save time only if you always made paté anyway. But if you never did before but do so now because you have a processor, you are spending *more* time in the kitchen, not less.

The difference between the practical and emotional responsibilities is this: If the husband does the laundry, he assumes the *practical* responsibility for that chore; but if the wife must remind him to do it or worries whether he's doing it well, she assumes the *emotional* responsibility for it. She is the one who *notices* what needs to be done, *remembers*, *reminds*, *initiates*, and finally *checks* to make sure it meets the agreed-upon standards. It comes down to an issue of trust: Will he do it in time, without prodding, and well enough? This must all be decided upon by the couple beforehand. The time you take in setting mutually agreeable expectations and standards will be the important component of how you both feel about the results.

Negotiating conflicts

Besides who does what on a routine basis, some of the other issues facing working couples deal with leisure time and vacations. Do they have different friends? Do they want to do more things together or have more private time for separate pursuits? Problems can occur simply in scheduling the little leisure time two-career couples have even if their friends, interests, and pursuits are the same. If needs are divergent, however, the problems escalate.

Companies are becoming increasingly sensitive to the needs of working couples. For example, they are more flexible about vacations. In the past, it was not as easy to take off from work at the same time. But that's not the only problem. Individual desires still have to be considered. Talking about meeting "her" needs on some occasion and "his" needs on others is important. We can help each other by knowing what it is that each prefers. Many people never even ask themselves the questions, "What do I really want?" and "What does my partner really want?" A vacation must be negotiated: "This year it's where *you* want to go; next year it's *my* turn to decide," or "The weekends are your preferences, but at Christmas, it's mine."

How to spend money is another important issue fraught with potential conflict. Should you spend it and enjoy life now, should you save it and enjoy life later, or should you save it for a rainy day? And whose money is spent for what? Should the accounts be separate or joint? In whose names are the charge accounts? Do you have equal access to the available money? Should the use of money be dictated by the amount each brings in? There seems to be no set rules. Each couple has to decide its own system. The point here, though, is that each couple *must* decide ahead of time how they will handle the money; it should not be left to happenstance.

Another conflict can arise from competition, especially when the two partners work in the same field or in the same organization. Since the tendency has been for women to enter the professional job market at an earlier age, many meet their future spouses at work and therefore marry men in the same field and with the same interests. A wife who makes more money or advances more rapidly than her husband may be a problem for the man, who has been socialized to see himself as the primary breadwinner. Careers can also collide if the two are employed in competing organizations, or if there is a conflict of interest. Even though many conflicts cannot be *resolved,* many can be *managed* if the couple is willing to share feelings and listen to each other with understanding.

However, the hardest of all issues for working couples to resolve is *mobility.* If one partner must relocate and the other has an equally good opportunity staying put, how does the couple decide? Most couples do not have good decision-making skills, nor are there role models from past precedents. Some couples decide to alternate cycles of opportunities: One partner's career can be the prime concern for a few years, and then the other partner's career has priority. The other spouse must always be taken into consideration, however, seeing to it that she or he is able to continue a career even if at a decelerated pace.

A new trend

One of the newer phenomena of our times is that married men now not only *expect* to have greater satisfaction from their work but to have available time for leisure in order to spend time with their families. This, of course, is the central issue for the multicommitted woman: She also wants both the opportunity for challenge and advancement and the time for other pursuits. A new benefit for a man with a working wife is that he can refuse a transfer, quit his job, change careers, or go back to school. The wife's job buys freedom for the man, which he previously did not have. Many are taking advantage of it: More men now risk changing jobs or careers to improve their income and/or the quality of their lives.[3]

A company that specializes in moving executives notes a new trend. Out of nearly 300,000 executives who will be asked to move to new locations this year, nearly *half* of them will object. A decade ago the refusal rate was no more than 10 percent. But is this due to the different values emerging for men in their 30s, or to the quality of life now seen as an entitlement, or to working wives?

Mobility decisions can be based on income, geographical location, or better career prospect for the other spouse. When mobility is an option for only one partner of a two-career couple, the problem should be shared with the organization involved. A boss should know the concerns of the employee, and may even be able to help by using personal contacts in the new location to find employment for the other partner.

Some couples justify short-term separations for long-term goals, such as splitting up during the week for work or school then reuniting on weekends. Others, having two jobs within a reasonable distance of each other, live in between and commute to work in opposite directions. One of my colleagues has a room on campus here at UNH; her husband, on a western Massachusetts campus. They have an apartment in Cambridge where they meet on weekends with the children, who are in boarding school. They say it works really well.

On the other hand, a friend of mine, 36 years old with a 4-year-old child, is director of a large project here in New Hampshire. She has a bright future, uses her skills well, and loves her community. Her husband, 37, has just been offered a lucrative position as regional head of a large company abroad. "A chance of a lifetime," he says. "I don't want to be a 'housefrau,' " she says. They might have to decide to live and work apart, with frequent visits between the two continents. A sad solution, you might think, but both would have been unhappy to sacrifice such opportunities at this time in their lives. She is not willing to risk her present position, nor he his future one.

These are the prices paid by the multicommitted. All a couple can do in the face of this conflict is spend the necessary time together; weigh all possibilities; alternate priorities in terms of each partner's aspirations, needs, opportunities for recognition, and advancement, and perhaps most important of all, self-fulfillment. With goodwill, almost all is possible! The secret is both love and *cooperation,* a willingness to compromise on the part of both individuals in today's world, with today's pressures. Love and cooperation are crucial to harmony *and* success. It demands a sense of responsibility, accountability, dependability, a yearning for self-development, and a *respect* for the growth of the partner.

The two-income couple as parents

All of the issues facing the two-income couple remain and new ones are added if this couple plans parenthood either immediately or ultimately. There are lifetime choices to be made: Have a child and interrupt the career? Have a child and continue the career? Delay childbearing?

Have a child now, work later (Delayed career) The question is how much later? Should a woman delay her career until the children are adolescents, or delay it only a couple of years until they are in nursery school? Either way she will enter the workforce later than her husband. When that time comes, he may already be settled in his career while she is only at the entry level or slightly higher, if she worked for a short period before having the child. Their career paths are beginning to diverge.

Work now, have a child later (Interrupted career) Again, the question is how much later? The biological time clock never stops ticking. A woman who delays her childbearing for a career has a limited amount of time to do so. When is the best time for this? When is it too late?

Some women elect to interrupt their work for only a relatively short maternity leave, staying at home less than a year. Other women stop for several years. In making a decision, it is important *not* to succumb to family pressures to take care of your child yourself, or to peer pressures to continue work. Be in touch with your own needs, hopes, strengths, and limitations. How long to stay home with a newborn will depend on your financial situation, your career aspirations, your enjoyment of your time at home, and your desire to get back to the office. What your friends are doing does not matter; there are fads about having large families one generation, small ones the next; staying home the last decade, going to work the next. There are no correct answers for all; do what is right for *you*.

Have both child and career now (Continuous career) One-third of all children under 3 years of age have working mothers. Less than 10 percent of these are in day-care centers. Six million children under the age of 6, and eighteen million between 6 and 14, have working mothers.[4] A woman who takes only a minimal leave when she gives birth and then resumes work has an integrated career and family life. Now, no matter what she chooses to do or how old she is, her concerns depend on the age of that *child*, not on her own age. Whether she is 20, 30, or 40 and working, her concerns vary with the child's age.

Most of the problems for working mothers stem from multicommitments. Her husband may be working his longest hours, in hot pursuit of his own career, while she is attempting to be Superwoman and to constantly juggle all her roles. She is trying to be the best wife, the most available mother, and the hardest worker, and feels

Age of Child	Concerns
Infant	Do I provide good enough substitute mothering? Is my baby okay?
Toddler	Is the person taking care of my child doing it the way I would?
Preschool	Are the day-care facilities adequate?
School	Are there after-school activities my child can join? Will she or he become a latch-key child?*
Adolescence	Does my child have enough supervision to keep out of trouble?
College age	Do I devote enough time to my child when she or he is home on school holidays?

*Latch-key children are those who carry their house keys around their necks so they won't lose them and are home alone after school until their parents return.

forever guilty for not accomplishing it *all*. This constant juggling takes an enormous amount of not only physical energy, but psychic energy as well. It can be exciting, depleting, challenging, frustrating, worthwhile, demoralizing, joyful, depressing, and totally fulfilling, all at the same time.

The issue is priority setting—choosing which responsibilities will be done less well or not done at all. Contrary to the old adage, I suggest the following advice: Not everything worth doing is worth doing well! However, priority setting is very difficult. Do you stay home with a sick child? Do you go to your child's play at school? Do you help out your needy parent? But what about the sales meeting, the convention, the report deadline? Where does the husband come in? Many companies still do not approve of their male employees' taking time off to care for a sick family member or attend to family business. Discuss these issues before they arise.

Day care

Child-care facilities are critical to a multicommitted woman, but they are still not as available as they could be. Even when good ones are available, nagging doubts and fears remain: How good are the day-care arrangements or the after-school activities? How good is the child-care person? How much can you *trust* that person?

Many European countries offer comprehensive day-care programs or incentives to mothers or fathers to stay home.* The United States has no comprehensive policy at all. Day care in the United States costs anywhere between $300 and $3,000 a year, and since most parents cannot afford it, they settle for neighbors, family, even unlicensed facilities. This problem will get worse as more women join the labor force.

Coping

The most critical skill needed by a multicommitted woman is the *coping* skill. There is so much to deal with. It is important to find time for both relaxation plus physical exercise. Frequently this is neglected because of the pressures of other demands from career, husband, children. The coping skill includes organizing time as well as knowing how to demand from others any help and support you need. Request assistance from your husband, children, parents; from your peers, bosses, friends; from anyone available and willing. Do *not* feel that you must do everything by yourself. Asking for appropriate help and support is one thing most women have not learned well. It must be learned early or women will end up feeling on a treadmill.

Time is the scarcest of all resources for working parents, so its use is an important issue. When the children are young, most things—short trips, sporting events, shopping, all types of leisure activities— are scheduled around them. As the children gain independence and go to school, the wife and husband become freer to spend more time together, but most couples still need to plan this togetherness in advance in

---

*Sweden offers fathers paid paternity leaves. According to *The Wall Street Journal* (1/29/79), men receive 90 percent of their salaries, drawn from a state insurance program, and the employers are required by law to hold their jobs open until they return. Although currently only 10 to 12 percent of the men take advantage of this, the number is rising. The average leave lasts forty-two days.

In Cuba, a law was passed in 1975 that husbands *must share equally* all household tasks with their wives. Although I do not know how enforceable it is, the premise here points to an equal distribution of responsibilities and hopefully an equal distribution of opportunities.

order to get the full benefits. After all, most free time must be used for other things: resting from hard and long days at work, housecleaning, marketing, cooking for the week, doing the laundry, attending to correspondence. What seems to get totally squeezed out of the picture is *private time,* time for yourself, time that is unstructured.

Ask yourself this: What is it that you are willing to do less well? that you can have done by someone else? that you can delay doing? This is not a decision to make by yourself; negotiate it with your partner.

As the children grow older, responsibilities are not really lessened, but the focus changes. Adolescents are not easy to live with; they are struggling with their own needs of dependence and independence. Attempting to fulfill one of these automatically means that you are *not* fulfilling the opposite one. Therefore, no matter what you do, you cannot possibly win. If you are aware of this, you might personalize somewhat less the effects of adolescent irrationalities, and it will not be as difficult to survive this time.

Children with both parents on a career path are fortunate to have these role models. They are then able to see at close range how men and women deal with life, with its opportunities and its pitfalls. In addition, parents who both work will tend to encourage their boys and girls to be androgynous, seeing to it that their children of both sexes have equal access to schooling, sports, and leisure activities. Many children of multicommitted couples develop independence, a sense of autonomy, and more self-reliance, and they go out into the world with this advantage.

I would like to share with you a quote from a former male colleague of mine:

> We have had—and continue to have— a quite extraordinary experience in marriage, not one we were perceptive enough to seek when we fell in love nor even one we welcomed fully when its depth first became manifest, but one with unique meaning for our own growth: the experience of living with a complete woman. We had been socialized to expect less and settle for the creature comforts that came with the acceptance. Along the way, we may have shown some lack of grace in surrendering the customary conveniences; yet in return we have been granted a love the richer and fuller for having its roots in mutual respect and growing maturity. That, we acknowledge, we owe to our wives. Had it not been for their capacity to insist on respect for their integrity, it is doubtful we would have achieved what we now enjoy. Of course it was not unilateral; we claim our due for what we have fashioned together; but we know it would not have happened without what they brought us: gentle courage, soft strength, selfless individuality.
>
> Our wives, our ''successful professional women,'' are our lovers, our friends, and our comrades in loves and lives of high adventure. We consider ourselves the most fortunate of men.[5]

LIVING INDEPENDENTLY

To remain single is an option taken by more and more women, as evidenced by the decreasing marriage rate and the gradual rise of the median age at the time of the first marriage. According to new census figures, the number of Americans living alone or with a partner, although not married, has risen more than 40 percent over the past six years—a rate of increase that is seven times faster than that of married couples starting new homes.[6]

A study of the relationship between the occupations and income levels of men and women and their marital status reported an important finding: Men in higher status occupations are *more* likely to be married than those in lower status occupations. For women, the situation was reversed: Those in higher status occupations are *less* likely to be married than those in lower status occupations.[7] There is one obvious reason for this. Higher status occupations tend to require more after-hours work. For men, being married is generally an advantage; wives usually take care of many details, such as laundry, cooking, and cleaning. For women, marriage can be a disadvantage, since it is difficult for them to do after-hours work as well as those chores wives usually do for their families. Again, what we see is the high price paid by professional women in terms of a personal life, for the rewards of a career.

Finances are often the major concern of single women. Compared to their male counterparts, the pay is lower and there are fewer opportunities for advancement. Working women earn 58 percent of the income earned by working men of comparable positions. The majority of women living alone or with unrelated people have incomes below $6,000 a year. Many of the youngest and oldest women of our population have incomes of less than $2,000 a year.

Although gains have been made in this area, there is still a long way to go. The Equal Pay Act of 1963 requires the same pay for men and women, and the Equal Credit Opportunity Act of 1975 protects women against credit discrimination. However, it is often too easy for credit institutions to get around these laws. Also, obtaining credit can often be difficult for the woman who is divorced or widowed. Most of these women have always depended on their husband's credit, so they have no credit rating of their own and are labeled a poor risk. Be sure to use your own first name on credit cards, not your husband's. You are Mary Smith, not Mrs. John Smith.

Women who need to obtain financial help or advice do have an alternative: They can go to a bank or credit agency operated for (and usually by) women. There are over 200 such consumer credit counseling service centers in the United States, and they will give free advice and help you to work out a budget, if this is what you need.

Single women are often faced with social pressures, especially on the job. Wives of male colleagues may see single women as a potential threat to their marriages, which naturally is not helpful in forming professional relationships with male co-workers. One woman whom I interviewed said that she often had to put up with "teasing" from the men in her department. She felt that may of these men seemed threatened by her choice to remain single, even though several of them were single themselves. At times, single women are more prey to sexual harassment than married women because assumptions are made about their availability.

The single woman composes a large part of the American work force. There are now more than 15 million female workers, many of whom have children. As of March 1975, the number of female-headed families totaled 7.2 million, which is one out of every eight families in the country.[8] The single largest issue facing single working mothers is the issue of child care. There is no husband to share that responsibility and the financial costs of facilities often are quite high. This often limits the single mother's choice of child care. Many employers are beginning to recognize and respond to the dual responsibilities of the single working mother. There has been more experimentation recently with flexible working hours, part-time work, and staggered work schedules. The results so far have been very positive, but we must keep exerting pressures on business to make these types of programs more common.

Instituting programs such as these, and expanding the pre-existing child-care programs may allow more women to enter and stay in the job market. Unfortunately, today's lack of quality child care and rigid work schedules have severely limited women's ability to fulfill both work and family responsibilities.

Often, because of moving to a different area when first entering the job market, the single woman may feel a lack of support. This is one of the first and most important strategies to think out. Some sources of support are other people on the job, friends, neighbors, family, local women's centers, evening college courses, and (for the single mother) groups such as Parents Without Partners. Men do not need to seek outside help. They find ready-made support groups in the camaraderie of their colleagues. Women must create a support group for themselves.

Although there are many problems for single women to surmount, there are also many joys. Many of the women living alone whom I have interviewed spoke of a sense of freedom. There are no constraints or restrictions on choosing friends, pursuing interests, working or spending leisure time without having to check with anyone, without being accountable to anyone. They spoke of feeling empowered by their ability to take care of themselves and, in some cases, of their children. Many said they were not willing to give up this freedom.

Living together

Living with someone when you're not married, although more and more acceptable in our society, can still represent a problem when working for a very conservative company. Do you bring your "friend," "significant other," "fiancé to the office parties, and introduce him to your boss and colleagues?* What is allowed in one place may not be okay in another. Ask a few close colleagues for their opinions.

A problem infrequently dealt with is the issue of homosexuality. Does a gay woman bring *her* female friend to office gatherings? This is a painful issue, for most places are intolerant of such relationships and would penalize the woman for her preferred life-style. On one hand it is wise to protect yourself; on the other hand, how will acceptance for differences grow if the differences are never exposed? My experience has been that a gay woman who has earned respect for her competencies and contributions may, once she is accepted—but only then—attempt to cautiously read some signs from close associates as to the wisdom of disclosing her life-style. For many it will never be possible. Managers should be sensitive and supportive of their subordinates with alternative life-styles.

---

*No one has as yet come up with a socially acceptable term to describe the partner of such a relationship. This can present problems at introduction times. The census bureau uses the term POSSLQ (Person of the Opposite Sex Sharing Living Quarters)!

Single working mothers

The balancing act performed by the single working mother is the hardest of all. Exhaustion is her constant companion. There is little time to take care of her own needs when there are financial pressures, career aspirations, and children who need mothering. To gain relief by finding some support is the only way to cope. Perhaps the working mother can find others in the same situation and they can share the burden of bringing up children alone.

LIVING DEPENDENTLY

The woman who is married, has children, and does not work outside the home follows the more traditional pattern of dependence upon her husband's income. This category also includes people who live together but are not married. Many women find complete satisfaction in the homemaker role; those who need to supplement family income or who experience restlessness as the children grow older may go into volunteer work or hold part-time jobs.

Recent societal pressures to relegate a "just housewives" status to women who work at home do these women an injustice! Being a homemaker is undervalued in this society. Even though most life choices are not without ambivalence, whatever choice we do make ought to be made with pride and with full commitment. The commitment could be more total, perhaps, if it were seen as having boundaries, a beginning and an end.

One of the main problems for women dependent on another's income is lack of preparation for later years. When the children have gone, these women often have no internal resources for either meaningful work or meaningful leisure activities. The average age of widowhood is now 56 (there are twelve million widows in the United States today); the midlife divorce rate is rising. How can all these unprepared women cope with living alone and being self-supportive without the necessary skills to do it adequately? The preventive measure is to start thinking about middle age early in life: Understand financial management, which is often done solely by the husband; take courses that might be useful later on; enjoy activities that have a future. In other words, actively plan for the later years throughout your lifetime. The chapter on retirement will deal more specifically with the older woman.

TAKING STOCK OF OUR AREAS OF CONTROL

As we look at life circumstances, we must remember that any kind of extreme—illness, death, personal tragedy, war—will disturb our adjustment. Over these we have no control. But we *do* have control in five areas, each with two parts. Every one of us should do a yearly stock-taking.

Five personal control areas

The first is the *physical* area:

- Am I exercising my body enough?
- Am I resting my body enough?

The second is the *mental* area:

- Am I learning enough?
- Am I teaching enough?

The third is the *emotional* area:

- Am I loving enough?
- Am I well-enough loved?

The fourth is the *social* area:

- How much in touch am I with myself?
- How much in touch am I with others?

The fifth is the *spiritual* area:

- How meaningful to me can I make my life?
- How significant to others can I make my life?

Women of today have many new opportunities opening up for them. New laws and increasing awareness of the working woman's role in society are helping women to overcome the many barriers that exist in the world of work. By looking at the problems and issues that women face, it may be easier to recognize them when we come across them, and once we recognize our barriers, we can begin to overcome them.

We, the multicommitted women of today, need to look at our lives in terms of costs and benefits. What and how much are we willing to give up in order to gain what and how much?

These are exciting times, with more possibilities than ever before for us to become whatever we decide, for us to live the lives we choose. The choices are available. The choices are ours.

NOTES

1. Vidal Clay, *Women, Menopause, and Middle Age* (Pittsburgh: Know Inc., 1977), p. 109.
2. J. Vanek, "Time Spent in Housework," *Scientific American*, November 1974, pp. 116–120.

3. Betty Roberts, *Middle-Aged Career Drop-Outs* (Cambridge, MA: Schenkman, 1979).

4. "The Early Years, The Middle Years, The Later Years," in *Women in the Labor Force* (Washington, D.C.: U.S. Department of Labor, Bureau of Labor Statistics), November 1975.

5. Theodore Nadelson and Leon Eisenberg, "The Successful Professional Woman: On Being Married to One," *American Journal of Psychiatry* 134-10(October 1977):1071–1076. Reprinted by permission.

6. "The way singles are changing the U.S.," *U.S. News & World Report,* January 31, 1977, pp. 59–60.

7. Elizabeth M. Havens, "Women, Work and Wedlock: A Note on Female Marital Patterns in the United States," *American Journal of Sociology* vol. 78(1978), pp. 975–981.

8. Patricia Hatiak, "Day Care Programs—The Unmet Neet," *Sojourner,* September 1978, p. 10.

SELECTED READINGS

Bird, Caroline. *The Two-Paycheck Marriage.* New York: Rawson, Wade, 1979.

DeFrank, Barbara. "Days of Our Married Lives." *Boston Globe,* July 11, 1978.

Greenleaf, Barbara Kaye, and Schaffer, Lewis A. *Help: A Handbook for Working Mothers.* New York: Crowell, 1978.

Hall, D. T. "A Model of Coping: The Role Behavior of College-Educated Women." *Administrative Science Quarterly,* vol. 52, 1972.

Hall, Francine S., and Hall, Douglas T. *The Two Career Couple.* Reading, MA: Addison-Wesley, 1979.

Klaus, Marshall, and Kennel, John. *Maternal-Infant Bonding.* St. Louis, MO: C.V. Mosby, Co., 1976.

Otto, Herbert, and Otto, Roberta. "The Juggling Act: Home and Career." In *New Life Options,* ed. by R.K. Loring and H.A. Otto. New York: McGraw-Hill, 1976.

Rapoport, Robert, and Rapoport, Rhona (Eds.). *Working Couples.* New York: Harper/Colophon, 1978.

Rowbotham, Sheila. *Woman's Consciousness, Man's World.* New York: Pelican Books, 1973.

Rowe, Mary P. "Prospects and Patterns for Men and Women at Work to Be Able Both to Learn and to Work." Centennial convocation, Massachusetts Institute of Technology, June 1973.

Scanzoni, John H. *Sex Roles, Life Styles and Childbearing: Changing Patterns in Marriage and the Family.* New York: The Free Press, 1975.

Shaevitz, Marjorie, and Shaevitz, Mortimer. *Making It Together.* Boston: Houghton Mifflin, 1980.

Chapter 8

CHOICES

Up front,

        or last row

Out there,

        or hidden

Speaking up for,
speaking out against,   or shutting up
        into safety

Be heard, visible,
criticized, attacked   Be quiet, unnoticed,
        left alone, peaceful

To shout the anger,
To cry the tears
To educate, correct,
    confront
        To mumble in corners,
        grumble with friends,
        protect myself from
           learning,
        protect others from
           changing

The Middle Manager: Managing Yourself

Only 6 percent of all middle management positions are held by women. Were you promoted because you did such a good job as supervisor? Were you hired at that level because of some particular expertise? Do you think you got here because of Affirmative Action requirements? And even if the latter is not true, is that what *others* are saying?

You now have fewer rungs above you; only directors, vice presidents, and the chief executive officer—the people who make policy decisions—are higher. And you have a couple of rungs below you—the supervisors, who carry out your directives and see to it that the job is performed adequately, and the workers who do the job. You are the person in the middle between the company heads and the rest of the organization. Besides the direct line relationships, which are vertical, there may be staff positions that relate horizontally to your unit. Budget, advertising, and personnel are staff positions that exist for the purpose of production and sales. In commercial organizations the point is to make money. In nonprofit organizations, public or private, the objective is to provide quality service, taking into consideration the cost-benefit factors. Those in a position to make decisions about how the money is to be made and the services are to be rendered are the people in line positions.

The supervisor teaches, trains, coaches, and gives advice. It is a prescriptive mode of relating to people, so you need to have *technical* expertise. As a middle manager, the focus becomes more social. A middle manager needs to know the concepts of organizational behavior, needs to understand the nature of groups as well as human nature. You will need to relate to people and events outside your organization, to your peers and superiors outside your department, and to your boss and your subordinates inside your unit. These various relationships need all your skill and diplomacy to maintain, to expand, or to let go.

For every competence that you possess, you need its flip side as well, in order to be effective: You must know your company's objectives but also know the limitations of its people and resources. You demand the highest quality but are flexible enough

141

A Manager	A Supervisor
Sets goals	Meets the goals
Plans	Implements the plan
Does not schedule others' work	Schedules workers
Anticipates problems	Solves the problems as they occur
Anticipates the staffing of supervisory positions	Hires workers as needed
Knows all relevant functional areas	Knows own area
Cannot replace an absent staff member	Can replace absent worker
Delegates training	Trains
Finds new resources	Distributes the resources
Represents the company	Represents the workers
Needs information from company executives	Needs information from immediate superior
Spends time with peers in other departments	Stays in own department
Sees people outside the organization (clients, suppliers)	Does not see people outside the organization
Mediates and negotiates at organizational levels	Mediates and negotiates at own unit level
Has ceremonial duties (speeches, etc.); is involved in community affairs	Has no ceremonial duties

to make adjustments. You represent your company, yet you represent yourself. You are always gathering information and are always sharing it. You are focused on results and are people-oriented. You need to listen and you need to speak out, and all of it at the same time, yet also in a sequence.

THE MANAGERIAL FERRIS WHEEL

The managerial ferris wheel never stops. There is no starting place, nor a place of arrival. It just keeps circling, with only the focus of attention shifting as your position changes. The higher you go, the more you will be involved in policy decisions. The

The Managerial Ferris Wheel

Gather information
- From people outside your organization (clients, suppliers, business and professional groups)
- Know the market
- Have lunches
- Make speeches
- Attend ceremonies

Share information
- With your superiors about your results
- Negotiate for more resources, new products, better services

Upper levels of management

Gather information
- From peers outside your department
- Socialize
- Arrange meetings

Evaluate
- Quality of product or service
- Give feedback to your staff
- Know the "bottom line"
- Make adjustments

Lower levels of management

Gather information
- From your superiors about organizational objectives and available resources
- Meet with directors, trustees, etc.

Share information
- With staff
- Allocate resources
- Hire
- Train
- Negotiate
- Delegate

lower you are, the more you will be involved with implementing the decisions of others.

The flow of the managerial ferris wheel goes from knowing what needs to be done, to seeing to it that it gets done, to evaluating how it was done in order to determine how it should be done better, faster, or just differently next time around.

The woman on the managerial ferris wheel is sometimes up on top, feeling a little heady, with a wonderfully clear view of the landscape; other times you are at the bottom, slightly sick to your stomach, not sure there is a way out. You are praised, you are blamed; people love you, people hate you; things are wonderful, things are terrible; you are excited and challenged, you are overworked and exhausted. What are the personal skills you need to cope with your new managerial position?

PERSONAL SKILLS

Before you can manage others, it is important to be able to manage yourself. There are five personal skills a manager needs to learn in order to manage herself effectively:

1. Self-knowledge

2. Risk taking

3. Decision making

4. Time management

5. Stress management

If you learn to know yourself well enough to trust your knowledge *and* your knowing, if you take the appropriate risks, make the correct decisions, manage your time effectively, and cope with the stresses produced by your responsibilities and by your visibility as a lone woman, you will be ready to manage others with competence and comfort.

Self-knowledge

Who you are as a person will influence how you will feel, act, and be as a manager. It is therefore important for you to sit down and take the time to think about three major areas: your goals (where you are going), your competencies (what you know and do), and your value system (what you believe is right).

Your goals are both personal and professional. What do you want to achieve? How far do you want to go? What do you want out of your professional life? What personal needs and satisfactions are you willing to sacrifice for your professional aspirations? Which professional objectives are you willing to sacrifice to achieve what you want in your personal life?

If you decide to sacrifice nothing and have both, you will sacrifice sleep, for sure! Your goals will determine how hard you work and the arena in which you will spend most of your time. Write down your personal and professional goals and note any discrepancies between the two.

Now make a list of your competencies and decide which strengths you are going to build on in your job and which weaknesses you must pay attention to.

How about your value system? Is it something that you have chosen for yourself, or is it something that has been imposed upon you? Either way, it is time to reassess the validity of your value system in your present life and see how it relates to your organization's values as expressed through its norms. Are there any incongruences between the norms of your company and your value system?

Being aware of the conflicts between personal and professional goals, using your strengths and knowing your weaknesses, understanding discrepancies between your personal value system and that represented by your company—all three form an important step toward building a concept of self, of who you are.

Risk taking

Knowing where and how to take risks is one of the most important skills a manager needs to have. Every time a manager hires someone, delegates a task, forecasts, approves a product, or suggests a course of action, she is taking a risk: The employee may not fit; the task may not be done well; the forecasting may prove wrong; the product may be defective; the action may be a poor move.

Risk taking is tied to self-esteem and self-confidence. If you are feeling good about yourself, you'll be able to take risks because you'll be more optimistic about their consequences. If you are feeling very tentative about yourself, you will be afraid to trust your own judgment.

Of course, what is a risk for you may not be a risk for someone else, and vice versa. You know when you are taking a risk by the signals your body gives you. I get a knot in the pit of my stomach and a tightness in my throat. If I'm standing with a paper in my hand, the paper will shake, so I try to place papers on solid surfaces. If I am sitting, I have learned not to lean forward, but to lean back so that I feel anchored against my seat. If I'm standing up, I try to walk back and forth and not look too paralyzed. Even though bodily cues will give off important data, it is cold hard facts plus intuition that will provide you with the best analysis of risk factors.

There are three major steps to risk taking. The first one is to *establish the desired outcome*: What is it that you want to see as a result of the risk taking? Some risks will allow you a measure of control, others will not; however, you have more control over the outcome if, instead of saying, for instance, "I want to get a raise," you say, "I will ask my boss for a raise on Monday." Mere wishing for a raise does not give you control over the outcome, you must rephrase it in terms of a positive action on your part.

The next step is to *predict the results* of the risk taking. You must decide who the risk taking will affect besides yourself. Will it affect anyone adversely? What is the worst that can happen if your risk taking does not work out, and what are the consequences of that for you and for others? What are the benefits if it works? This will help you assess the situation realistically.

The last step is to *actually take the risk*. You take the necessary steps to accomplish the desired outcome, evaluating all along the way whether your predictions of the consequences were correct.

As an example, let us say you need a larger budget to start a new project. Your desired outcome is a budgetary increase for a specific project. You have decided that you will go to your boss next week and ask for that increase. The strategy would be to first send a memo saying that you wish to talk about the project, or else be prepared to broach the subject in person. If you do the latter, be sure to make it clear that a final decision right then and there is not necessary. Many people, when taken by surprise, will start by saying, "No." Next, make a list for yourself of all the reasons how this project will benefit your boss, your department, or the company. Then predict the outcome. The best thing that can happen is that you get your budget; the worst that you do not or that you will be perceived unfavorably for asking. Most people will respect employees who have creative ideas they can back up with evidence of having thought them through. If *you* don't feel that your project is worthwhile, you won't get your *boss* to think so either.

It is frequently difficult for women to ask for increased funds; asking for money has an unpleasant ring to it. Yet this *is* what needs to be done. Put yourself in your boss's place. If you have a certain budget to distribute, wouldn't you give it to the people who make the most demands rather than to the quiet ones who seem to be satisfied with whatever they have? What women need to learn to do is to make their needs known and ask for their demands to be met.

Risk taking can take many forms. It can be in terms of hiring an employee, of financing a certain project, of investing in equipment, of accepting a customer's deadline. The managerial road is paved at every turn with decisions, and each decision is a risk. By analyzing the desired outcomes, predicting the possible negative and positive consequences, and delineating the action steps, a manager can minimize the risks of risk taking.

Decision making

Most people are not promoted to middle management unless they are perceived not only as being decisive but also as being able to make quick decisions in the face of ambiguity and inadequate data.

I would like to propose now that you not see yourself as either a risk taker or a non-risk taker, as either welcoming the challenge of the unknown or needing the security of the known. Instead, take this aspect of your personality as only one of the factors in the decision-making process. For after all, risk taking is making a decision—the decision to take, or not to take, a risk. *Every* decision has this risk component: It may be a poor decision with negative consequences. The results of a poor decision may vary anywhere from going unnoticed by a majority of people, to making you look incompetent, to having the company go bankrupt, if you are in position to make those types of decisions. However, whether the decision is to cancel a class because of the bad snow storm, as I've done on occasion, or to rethink a total course curriculum, the process is very similar.

Is the decision-making process different for women than it is for men? I believe it is for the women who have been socialized according to the saying: "Man proposes, but woman disposes."* Even though a woman makes a decision, it is at the man's initiative. In other words, *he* offers the options from which *she* can choose. Think of the whole dating process: *He* must ask first so that *she* can say "Yes" or "No." Even though this may sound a bit old-fashioned today, the asking for the hand in marriage is the same thing: A man offers the option; only then can the woman choose. There is evidence that these customs are in the process of changing. However, just the other day I went to a restaurant where the men were given menus with prices, whereas the women were given the same menus without prices. This allowed the men to make *informed* choices; the women made *un*informed ones.

*An old French proverb, "L'homme propose mais la femme dispose."

The steps to take When is a decision needed? When there is either more than one alternative available for a behavior or an attitude, or when there is the perception of disequilibrium. In either case, there are several steps to follow that could help you in this process.

One: Determine whether you alone will solve the problem or whether you will request participation from others. Base this decision on two things: (a) Will others be affected by this decision? (b) Will they be committed to its implementation if they've had no say? Frequently, a manager must decide alone anyway—for instance, when an immediate decision needs to be made and there is not time to consult anyone or get more information, or when a participative approach will be so time-consuming that it will be unproductive, or when the manager predicts irreconcilable differences among her colleagues or subordinates and prefers *imposing* a decision on the group. An example of this last situation would be needing to fill a given slot from an available pool of equally qualified people. The manager can and should inform the people concerned of the reasons for her choice, but she cannot abdicate the final decision to others.

Two: Gather all the information you can. See that you obtain all the required information from whomever has it. You can share your problem with others *individually* and get their ideas and suggestions, or you can get these people in a group to hear their ideas. After you have the data, either you alone can make the final decision, or you can act as a chairperson with the group to make a joint decision. If you opt for that method, you must be willing to accept and implement whatever solution the group comes up with. To not make a decision or to delay a decision is in itself a decision with particular consequences. If, for example, you feel you need more information before you can decide, you have to weigh the cost of the delay against the quality of the eventual decision.

Three: Define the problem in terms of whom it will affect now and whom it will affect later. List your gut feelings about it, how it fits your value system, your personal hunches about it, your preferences for the outcome. Set these thoughts aside as important data. Make a list of the organizational or group objectives in terms of the desired and predicted outcomes of the decision. Now check to see if there is a conflict between your personal list and the organizational list.

Four: Generate alternatives—as many as you can—in a brainstorming session, either alone or with others. This will avoid the trap of an either/or way of thinking. There is seldom a *best* decision, but usually a *better* decision.

Five: Pick the most appealing alternatives and look at each one in terms of possible consequences, both short- and long-term, for you, for others, and for the organization. Do not forget that consequences usually have side effects and screen for those. Also, ask yourself if one of the solutions to the problem might create new problems, and if so, what they might be?

Six: Make the choice, keeping in mind that (a) it should be acceptable to you and to those affected; (b) it must be possible to implement with currently available resources; and (c) you should feel that you can live with the possible risks this decision entails. To figure these out, ask yourself what is the worst that can happen.

Seven: Plan the implementations. How you implement the decision is as critical as the decision itself. Are you going to handle it yourself? Will someone else do it? Is the decision going to be announced at a meeting, in front of a large group, individually, or by written memo? Do you plan to share the steps you took to arrive at this decision? Are you open to criticism? Are you willing to change some aspect of your decision if provided with new information and/or feedback?

Eight: Check results. How soon do you plan to follow up on the implementation of your decision? How frequently will you need to check? What will you be looking for? If you get feedback, do you have a process by which a review can improve the quality of your initial decision?

This sounds like a very cumbersome process. If you do indeed have a difficult decision to make, I urge you to go through these steps. After you have gone through the process several times, it will become familiar to you. Then you will be able to do this much more rapidly, having integrated its principles so they become part of your managerial style.

Getting information   Most managerial decisions are not based solely on a systematic and logical process of data gathering, but also on an intuitive sense of what will work. This sense comes from the knowing part of ourselves, which we need to learn to trust. However, many decisions that involve long-range planning must be of high quality and based on sufficient solid information. To obtain enough good information to make a quality decision, a woman needs to have a network.

As we have seen, it is difficult for many women to be part of the informal male network, where critical information is shared. Without this access, she will have missing pieces. If a woman manager feels that she does *not* have all the relevant information, she has little choice but to share her decision-making process with the people who *do* have the information. After all, she will be held accountable for the results, which may be less satisfactory if exclusion from the network of information is not overcome. Good decisions may be made sitting alone at your desk behind closed doors, but only after you have frequently left your office to gather information. And you should leave the door of your office open for people to come in and share with you whatever you need to know.

Ask questions—lots of questions, from lots of people—above you, below you, at your level. Stop a moment and determine which people in your organization are likely to have the information you need. How can you get it from them? Which people already give you information on a regular basis? Is there a difference between those two groups? What do you need to do to expand your network? One of the ways to enlarge your information gathering is to provide information to others. Make a list of the people to whom you give information and those who ask you for information. Do you see a discrepancy between the number of people who come to you and the number of people you seek? To *get* information, you need to *give* it; to give it, you need to have it; to have it, you need to get it: It is a never-ending spiral.

Another issue that differentiates men from women in the decision-making process is in the fact that so many women are tokens in their organizations. In this case, a poor decision from a woman will stereotype women as bad decision makers. No one will say, "Joe made a bad decision. Well, no wonder, he's a man!" But you might hear, "Mary really goofed! Well, what do you expect from a woman!" Because of her higher visibility, a woman risking what seems to be exactly the same as a man is, in fact, risking more. If more people look, more people see, and whatever a woman does takes on added significance. A woman's mistakes are less apt to remain anonymous or unnoticed. This is tough! After you have taken a risk that has not paid off, all you can do is what the men do: shrug your shoulders and say "Win some, lose some!" But in *no way* should you allow it to cramp your style the next time around.

Men know how to make a joke out of their mistakes. Women, on the other hand, tend to personalize failure, a handicap that blocks us from trying again. If I make a mistake and feel like a total failure, I will be less likely to risk doing it in the future. Let us also learn to laugh at ourselves. People respect men *and* women who can make a mistake, learn from it, and keep going. If you consider a mistake as just one wrong move among the many right moves you have made in your life and still plan to make—if you remember that you never make *the* wrong decision but only *a* wrong decision—it will be easier for you to risk decision making and to decide to risk.

Time management

Studies show that managers seem to control no more than 25 percent of their own time, and the remaining 75 percent is not under their own discretion. The most precious resource we have is time. Free time is made, not found; we need to learn to "make time" for ourselves and "use time" effectively with others.

You barely have time to turn around before the hours pass, the day is gone, the year is spent, and life is over. Let us all stop and reconsider. A day is 1/6000 of the next quarter century; that is, we have 6,000 days for the next twenty-five years. Therefore, today should, in some way, make 1/6000 of a step toward an ultimate goal. The way to get in touch with such a distant goal is to think in terms of an obituary: If someone were to write one paragraph about you and your life twenty-five years from now, what would you want written? How would you like people to remember you? List some of your personal and professional accomplishments and some of the qualities you are proud to have. As I write this, I am thinking about what I would like to have accomplished in twenty-five years and how I would want to be remembered. I would like to be remembered for having empowered women. What about you? To manage your life you need to manage your time.

Prioritizing goals   Good time management requires setting priorities according to a personal life plan. First, identify your life goals; second, list the stepping stones to those goals. Decide where you need to be five years from now. Ascertain what you need to know or do in order to get there. What would be your six-months' objective? What should you be doing next month or next week to achieve this? What are you going to do about it today?

Alan Lakein in his book, *How to get Control of Your Time and Your Life* (Signet, 1973), lists several ideas that I believe are very helpful. He suggests that you make a list of things to do if you knew that you would be struck by lightning six months from now. The purpose of this is to get in touch with the things that are really important to you that you are not doing, and to see if there is room for them in your life at some point in the not-too-far-distant future.

To achieve your goals, you must work toward them systematically. Make a list of *all* your goals. Once goals have been written down they must be *prioritized*. Assign each goal an A, B, or C, then further prioritize the A goals into A-1 goals, A-2 goals, etc. The point is that each goal must be assigned a priority. A-1 goals are those with the greatest personal and organizational payoff. Occasionally your bosses' A-1 goal must of necessity become our own A-1 goal. C goals are of little value, and should be questioned as valid goals. Once the goals have been prioritized, action plans, consisting of the steps necessary to achieve the goal, can be developed. It is helpful to be as detailed as possible in determining an action plan, since this will ultimately make the goal easier to accomplish: You take many small bites rather than one big gulp.

Next, Lakein suggests, start each day alone in your office, for as long as it takes, to write a daily "to do" list. Base this list on objectives you need to accomplish for your work, whether for this particular day or week, or for the month or even the year. Remember that daily objectives will encroach upon long-term goals and always diminish that list. Prioritize each item on the "to do" list just as you did your goals. Priorities should be up-graded and down-graded continuously, and items that retain low priority should be dropped after a while, if possible.

Even though some of the high-priority items may take longer, and you would prefer doing the things that can be expedited in a few minutes, it is usually not productive to wait for a long block of time in order to accomplish a particular piece of work. That long block of time very seldom comes, so you are better off taking that item and dividing it into sections in order to accomplish it in shorter periods. Be sure to have on your list only the things that have high value for your personal and professional goals, and those of your organization.

How much of your time do *you* control? How much do the *pressures of your work* control you? Ask yourself the following question several times a day: "What is the best use of my time right now?" The first answer to this question is another question, "Do *I* need to do this or can anyone else do it?"

Time robbers What are the most common sources of problems in time management? They seem to be recurrent themes, shared by most managers. They are:

1. Telephone calls

2. Meetings

3. Reports

4. Visitors

5. Not delegating tasks

6. Procrastinating

7. Firefighting

8. Special requests

9. Delays

10. Reading

Item one on the list is the *telephone.* If it is possible, have your telephone calls screened by a secretary. If you must answer them yourself remain brief. Otherwise, let the phone ring and do not answer, or take it off the hook altogether. Perhaps it is possible to have a telephone answering machine delay the calls by taping messages. When I am doing something that requires concentration and a telephone call interrupts my train of thought, it takes me a long time to get back to where I was (I have a long start-up time). For instance, as I write this the telephone is off the hook. When you return your phone calls, do it in a specific time slot, perhaps when you are least productive. For me, it would be right after lunch, when I am somewhat sleepy and not very creative. Most women traditionally have made good use of the telephone to meet their affiliation needs. It is a contact with others, a way of reaching out and being reached, a way of sharing. But in business situations, some women have trouble cutting the conversation short because it is a rejecting gesture. A short explanation such as "I have another appointment waiting," is sufficient in order to be able to hang up. I also have often said, "We can continue this some other time," but usually there is nothing that important to continue.

Meetings are another big time robber. Are there too many to attend? Are they too long? If you requested the meeting, be sure to send an agenda ahead of time to all those you want to attend. If you expect specific contributions from individual members, state so. Be at all meetings on time and let people know when the meeting is supposed to end, then stick to it. Meetings that you do not run can be handled in two ways. You can either find out if there is an agenda and what it is so that you can come prepared, or you can view them as a ritualistic observance of your company's norms. Often, meetings are held simply to get together in a predictable pattern; very little gets accomplished but the people's needs to be with each other, to share some information, and to recommit themselves to company goals is the unstated objective. Do not discount such ritualistic group meetings; however, they should not be overly long or overly frequent.

For women, meetings present specific problems. We need the visibility and we need to participate. We also need to call them so that we can be seen "in charge." The meetings must, of course, have a legitimate purpose. Although meetings are second only to phone calls as time wasters, do not shy away from calling them when needed, and on your own turf whenever possible. Meetings are also a place where we can become known provided we voice our opinions. But be sure you control them, if you're in charge, or control your participation, if you're not.

Reports that need to be written, that need to be read—how many of these are necessary? Once in a while you should re-evaluate the need for both the written input and output. If reports or memos need answering, comment right on the papers and return them immediately; upon reading, make notes while the subject is in your mind. Never let a piece of paper go through your hands more than once. It is now appropriate procedure to answer letters by writing a note on the sender's letter and returning it, with a copy for your files if necessary. Writing memos and reports gives women needed visibility, and making notes and comments on other people's reports are equally valuable in building your network. Reports fill a different need for women than they do for men.

Some *visitors* cannot be avoided; others can be seen during nonproductive times of the day or bunched together in one afternoon reserved just for visitors. This is basically the same problem as the telephone. Many women have difficulty in cutting short a visit, in refusing to see a visitor who comes at an inappropriate time, in making delineations between social and professional visits. Even if the visitor comes for a professional purpose, how much social interaction you permit or even encourage will depend on the importance of this particular visitor, the potential of this visitor for meeting your own goals, and the kind of information this visitor may have, if not now, possibly in the future. In other words, your time spent with visitors is strategic. What looks like wasted time now may fulfill some potential long-range objective. If in doubt, be nice, but be nice briefly.

What tasks can be *delegated* and to whom? This is one of the more important questions to be answered for managers. Many women have particular problems in this area. If we have a small territory, we tend to hold onto it, rather than having the foresight to train someone for our job so that we can go on to a higher one. The more other people can do things for you, the more you can do things for yourself. It might even be worthwhile to *train* someone to perform some of your tasks in order to free yourself for achieving your A goals.

To help yourself stop *procrastinating* make a list of the things you need to do that you like, and a list of those that you don't like. Usually, we tend to delay doing what we do not like. If you prioritize the list, then you will do the A items first, whether you like them or not. Procrastination can be costly, especially if what you plan to do cannot be done at the last minute because of intervening events. Your delays can frustrate others, which means the cost can be in terms of goodwill. A procrastinator is *not* a credible manager. Do women procrastinate differently from men? I do not know of any research on this topic; however, I would guess that since many women are perfectionists, they might tend to procrastinate rather than turn in something they see as inadequate.

Even though *firefighting* is always an A-1 priority, a large number of emergencies may be an indication of poor planning. When they do occur, ask yourself if they were predictable and what you could have done to avoid them. Managers who firefight most of the time are managers who have not been able to avoid the fires, have not been able to see them coming, and have not prepared for them adequately. I believe women are especially good at sensing potential trouble. Women have good antennas and dislike emergencies, so most women managers should not be excessively involved in firefighting.

Everyone wants to be nice and accede to *special requests*. Women are perhaps especially prone to agreeing to do things outside of the job. "Won't you please do this for me?" "Won't you please help me with this?" "Just this once?" "It would be a big favor to me." "I'd be *so* grateful." "Thank God you're here, I knew I could count on you!" Who can resist these? Well, most men can, and most women cannot. As daughters, wives, mothers, and friends we have learned to accede to other people's requests. We have been socialized to be nice, helpful, nurturant; now, as a manager, to say "No" is a difficult behavior to learn. The decision must be based on what acceding to the request will do for you, for the other person, and for the organization. It might be that building or maintaining the relationship is the important factor; it might be that you are indeed the only person who can help. But it also might be that you just don't have the heart to say "No." *That* is not a good enough reason. Perhaps you can help these other persons find another resource, or let them do it themselves with some encouragement. Any categorical "No" should be accompanied with an explanation so that no one feels rejected. The point is not that these people are not worth your time; the point is that your time is worth managing.

Delays are often not within our control. However, when something is needed within a certain time span, one way to deal with delays is to either have your secretary check on it or check on it yourself. This will avoid being caught unaware. Often it is known when delays are forthcoming, and they can be damaging to any sequencing in which you are involved. It is legitimate to get upset at unexpected delays, especially if you suspect that the work *is* getting done for your male colleagues. It is not infrequent that a woman's work gets done *after* men's work, by secretaries or other people involved in fulfilling these tasks. Make your demands clear: State when you want the work finished and ask for what purpose or when the importance is established. If you sense unfair prioritizing, check on it and complain.

Tenth on the list is *reading*. This involves reading junk mail and whatever else is unrelated to your work. To use your time effectively, it is important to learn to scan material for key words. In this way you can see if it contains any information you may need. Scanning takes little concentration; it can be done in a relaxed fashion, sitting with a cup of coffee and your feet up, at a time when you are least creative.

Time savers What are some of the possible time savers? Using a dictaphone is one. In fact, if you drive to work, using a dictaphone in the car is a good idea. Others are putting time limits on everything and sticking to them; acting and initiating instead of only *re*acting and *re*sponding; consulting others; and my favorite—using the helper/helpee relationship, which is useful when someone requests help.*

It is not easy to be helpful effectively. The usual tendency is for the *helper* to define the help needed, to provide it, and to control that process. Ideally, the helpee should define the help needed and control the process. The manager should ask, "How can I be most helpful to you during this period?" The responsibility of the helpee is to define both the problem and the role that the helper is to play.

*Adapted from John Haskell, Director of the Center for Industrial and Institutional Development, University of New Hampshire.

Even though I have placed time management under "Managing Yourself," time does not concern yourself alone. We give time to others and take it from others. So time management has two parts: your own personal time and how you manage it; and your time spent with colleagues, bosses, subordinates, people outside of your organization, family, friends—all those who demand your time and have a right to it and those who want it but do not have a right to it. It involves your decision as to how you will dispense it—how much and of what quality, for you can be with people and give them only peripheral attention.

Time management is a particular problem for women because of the many demands placed on us by all the people in our lives. How much of our time does in fact belong to others? How much of it are we going to keep for ourselves? Every day we must decide how we will deal with this time in our lives.

Stress management

Depending on how hard you must work, how much you worry, how many commitments you carry—especially how many you don't carry off—you will experience stress. But just as risk taking is different for different people, what stresses you may not stress someone else. In addition, the amount of stress people can take differs. As women strive not only to get into the world of work but also to gain recognition and achieve success, stress becomes a constant companion. Success, with its components of mastery, aggression, competition, speed, constant flux, and unidirectional focus, produces stress.

Henry Mintzberg speaks of three characteristics of a manager's job: brevity, variety, and discontinuity.[1] These factors do not promote security, predictability, and peace. Various studies indicate that about half of all managerial activities last less than nine minutes. Only 10 percent exceed one hour. About 80 percent of a manager's time is spent in oral communications, and 93 percent of these are on an ad-hoc basis.[2] Is it any wonder that so many managers feel stressed so much of the time!

A 1978 study of hospital management found the following to be the most stressful events: (1) major changes in policies and procedures; (2) requirements to work more hours than usual; (3) an increase in the pace of work; (4) a new supervisor; (5) new subordinates; (6) a major reorganization of the department; (7) a change in the nature of the job; (8) a new co-worker.[3] If you notice, all these events include some type of change. So to reduce stress, try not to change too many things at the same time.

The multicommitted woman seldom has real choices in fulfilling her many roles. Stress is part and parcel of every working woman's life, and it can produce anxiety or depression. Anxiety is a feeling that something unpleasant or threatening is about to happen, even though there is no apparent cause for worry. It is related to fear, but while fear is a short-lived emotional response to a definite danger, anxiety is more vague and lasts longer. A person who feels depressed cannot accept help easily, is lonely much of the time, performs poorly even in areas of competence, and is blind to the good things in life. Depression is related to grief. But while grief is a reaction to a real loss, depression is a generalized feeling, often without obvious cause.

Some people seem to enjoy working under stress, and perhaps for them stress is not harmful. However, for many, stress leads to heart attacks, ulcers, hypertension, and emotional breakdowns. Stress is often caused by simple overload—trying to do several things at once to "save time." Because so many women start their careers late, they feel in a rush to catch up. I know I often feel this way. The question to ask yourself is, "Catch up with whom and why?"

As we rush through life, the details of it blur. Because I want to get this chapter done today, I will not take the time to sip a cup of hot tea and watch the river under my window. If I beat my own deadline and get an ulcer, will I be a grander person? Who will appreciate me more? How important it is to stop once in a while and just enjoy the moment! I don't know where the compulsion of getting things done comes from, but I *do* know that I respond to this inner pressure to do it. I say "Tomorrow I will have time for myself, for other things." It's never *today*. Ask yourself this question: "How high up is the top for me?" There is a myth that the hard worker, the person who works long hours, stays overtime, comes early, leaves late, and works on weekends, always gets ahead. It's not necessarily true. A great performance is important, but a good career is more so. Besides, upward mobility is not the only sign of success. A good *fit* with the job is. Growth is personal, and success is achieving your own potential, not meeting others' expectations. However, stress that ends with success is more tolerable than stress that ends in failure.

Stress can be the result of too much work, but it can also be the result of too little of it. Boredom can be as stressful as too much challenge. Women are often under stress because of both too much and too little. "Too much" is the constant reminder that she is different, the constant juggling of her many commitments. "Too little" is the lack of support from her male colleagues and the more routine tasks assigned to her.

It is more stressful to be in a minority, and in middle management positions, women are a minority. Because of this we stand out, which renders us visible, and visibility exaggerates the importance of all of our actions. To be constantly noticed is stressful, yet it is important to be noticed without being notorious. In addition, most token women are stereotyped and seen as speaking for all women or are discounted because of their sex. Assumptions are made about their abilities, career aspirations, and relationships with male colleages. Fighting these patterns of discrimination is exhausting and stressful.

A woman, especially a token woman, in a predominantly male work environment has not only to learn new roles, but must constantly be faced with people who question her very existence in the organization. Feeling tentative in an environment is stressful. Living without self-confidence is stressful. What can a women do to help herself?

Coping with stress   There are three main coping strategies. One—*use support systems.* You must have a place where you can relax and some people with whom you can talk about your stress. Two—try to *organize your life* in such a way that some of the extra burdens that stress you are minimized. It is a matter of setting priorities. Trying to be everywhere at the same time is not necessarily commendable. (I am saying this to myself as much as to anyone else.) Three—*work on self-management*. What does that entail? There are five basic components.

Exercise. It does not need to be more than ten to twenty minutes a day. I exercise the three major parts of my body. I step about an arm's length away from a door or wall and do push-aways while standing up until I get very tired in my arms and shoulders. Then I sit on the floor with my knees bent and slowly let my torso go down to the floor, which is good for my stomach. Then I stand on the edge of a stair step and move up and down, getting my heels below the step to stretch the backs of my legs. I also make windmills out of my arms and turn my head round and round. All of these can be done in an office any time of day. You can also try sitting in a chair and holding your legs up as long as you can. In the morning before breakfast I put on some fast dance music and run around the living room until I get out of breath. It is not so crazy—you'll see how much music helps.

A balanced diet. Eat whole grains, lots of vegetables, and (as recent research seems to point out) less red meat and more fish and chicken.

Time management. As mentioned before, set priorities, making sure to do what you need to for yourself. Do not simply respond to pressures from others.

Quiet time. You need this to regroup yourself, to decompress, to relax. Meditate, take a nap or a walk, day dream, look out the window.

Support systems. Friends, family, peers, even certain associates—they're all necessary. You need people with whom you can discuss personal and professional goals, problems, strategies, and plans, and more important still, with whom you can have fun.

What are the characteristics of people who manage their stresses better? (1) They know themselves, their strengths, their weaknesses. (2) They have a variety of interests, so that they have many sources of satisfaction. (3) They react to stress in different ways, not always with the same pattern. (4) They acknowledge and accept the facts that people are different and that they themselves can change. (5) They are active, productive, and enjoy what they do most of the time.[4]

I once saw this headline in *The Wall Street Journal:*

WHAT PRICE GLORY?
FOR HIGH-LEVEL AIDES IN WASHINGTON
IT IS ALL WORK, NO FAMILY—
MARRIAGES OF SOME SUFFER. . .

The question is, of course, is it worth the price? And yet, it is all easier for men than it is for women, because even though those high-level aides have no time for their families, there is a *wife* at home taking care of family needs. As a woman, you have all the pressures of work, your responsibilities for the family (even when shared), and strains encountered because of your gender in the world of work. The stresses are enormous, perhaps greater than they have ever been before because of our own self-imposed expectations. What price Superwoman?

Women as a rule are more aware than men of what they give up by being totally focused upon their careers at the exclusion of other interests. There is real loss at giving up non-work-related pursuits, for women know the gratification of a multifaceted life, the pleasures found in family and friends.

Up to now this book has been a push to go, and now there is a pull to stop, or at least to slow down and take stock. A balanced life is a push-pull. As soon as the push gets too intense, there needs to be a pulling-in. When that tends to become too quiet, you need to move on. What is your own optimum amount of stress for maximum efficiency and happiness? You need to find this out and try to live accordingly. If you have made it this far, what do you still need to prove to anyone, including yourself? What are your pleasures; what are your pains?

Ask yourself if you really want the stress that goes with the job. If not, upward mobility may not be appropriate. Horizontal enrichment may be satisfactory—it can permit you to get involved in a variety of activities and be content with your job. On the other hand, if the top still beckons, then deploy all your forces to get there, but manage yourself intelligently on the way and care about yourself.

NOTES

1. Henry Mintzberg, "The Manager's Job: Folklore and Fact," *Harvard Business Review,* July–August 1975.

2. *Ibid.*

3. John D. Adams, "Improving Stress Management," *Social Change* 8–4(1978).

4. *Ibid.*

SELECTED READINGS

Adams, John; Hays, John; Hopson, Barry. *Transition: Understanding and Managing Personal Change.* London: Allenheld, Osmun & Co., 1977.

Bem, S. L., and Bem, D. J. "Case Study of Non-Conscious Ideology: Training the Woman to Know Her Place." In *Beliefs, Attitudes and Human Affairs,* ed. by D. J. Bem. New York: Brooks/Cole, 1971.

Benson, H. *The Relaxation Response.* New York: William Morrow, 1975.

Business Bulletin, *Wall Street Journal,* March 8, 1979.

Dittrich, Louise Cahill. "Modern Stress: Strategems for Self-Awareness." Unpublished paper. Boston: Harvard Business School Case Clearing House, 1975.

Heinen, J. Stephen; McGlauchlin, Dorothy; Legeros, Constance; Freeman, Jean. "Developing the Woman Manager." *Personnel Journal,* May 1975.

Hermon, Jeanne, and Gyllstrom, Karen. "Working Men and Women: Inter- and Intra-Role Conflict." *Psychology of Women,* Summer 1977.

Lakein, Alan. *How to Get Control of Your Time and Your Life.* New York: Peter W. Wyden, 1973.

Quinn, Jane Bryant. "A Woman's Place." *Newsweek,* February 26, 1979.

Ullyot, Joan. *Women's Running.* Mountainview, CA: World Publications, 1976.

Chapter 9

TRUST

I don't know
What is best for you.
But I trust you to know that
I know
What is best for me,
And I trust you
To trust me.

The Middle Manager: You and Others

There you are, sitting in the middle of the pyramid! You have at least one boss, who may or may not also have a boss, and you have subordinates, who supervise workers. There may be even further chains of command up and down the line on either side of you. But *you* sit in the middle. Your work has two main aspects to it: responsibility and accountability. You are responsible for your work, for your supervisors' work, for all the other managers below you who report to you, and you are accountable for the results of your management to your boss or bosses. You need to *know* on one hand, and you need to *tell* on the other.

THE ONE-UP ONE-DOWN RELATIONSHIP

All superior-subordinate relationships have a one-up and one-down configuration. However, this relationship may be based on *real* power or on only *attributed* power. When a person above you has defacto formal powers over you, such as having the power to decide whether to promote you, give you salary raises, and increase your area of responsibility, or when that person has a special area of expertise, the power is legitimate and is usually accompaned by a title and general public acknowledgment of the dominant role of this person. Traditionally, the white male has always been in the one-up position.

The one-up one-down relationship is based on attributed power when the person's role is not formalized by title or position and not legitimized by expertise, seniority, or age, but is simply a function of discrimination due to socialization. For instance, whoever is doing the "psyching out" is in the one-down position. "Psyching out" is going around with big antennas trying to pick up all available cues in order to behave in a manner that is pleasing to the other person. For this reason, I call one-down people "antenna carriers." They are the people who walk into a room and immediately try to figure out who there is important, and who might be threatening.

159

They are anxious to behave appropriately, but always according to someone else's criteria. The antenna carriers of the world need to psych out in order to survive. They are the women in most cultures; they are the blacks and other racial minorities in white-dominated societies; they are the lower classes; they are the ethnic minorities; they are the disadvantaged, the fearful, the subordinates, the servants, the children, the tokens, the foreigners. Many of these people scan their environment constantly to pick up any cues that might indicate potential trouble. They look for reactions when they test out new behaviors, and are willing to immediately change course if they sense disapproval. The one-up one-down relationship under these circumstances is only in the ingrained habits of people, in the *attributed* status of people. Status is in the entitlement inherent to the one-up position; it is in the eyes of the beholder in the one-down position. For example, most men generally expect to be understood; most women expect to do the understanding. Men expect to be catered to and cared for; women expect to subordinate their own needs to those of men. Men expect to be noticed and approved; women expect to do the admiring and applauding.

Most people in the one-up position feel free and satisfied in that position. Their needs are largely met by the one-down persons and they experience no need to share their feelings, opinions, or thoughts with one-down people. The one-down people withhold their feelings and thoughts because they perceive a high risk in sharing them. They thus deprive the people on top of needed information as to what is *really* going on, and the person who is cut off from truthful contact with subordinates will not be able to make the best decisions.

In the meantime, the one-down people spend a lot of energy psyching out the one-up people and taking care of their perceived needs, at least enough to survive in the relationship. One of two things happens. They either lose track of their own needs and subordinate these to the one-up people, or they are aware that their own needs are not being met and become depressed, resentful, frustrated, or angry. This will lead them to seek support from same sex, same race, or same class. These support groups can function for emotional support, or they can become political coalitions, such as the Women's Movement, black caucuses, and unions.

There is a strategy women ought to know about—the strategy of mystery. A person who shares very little personal data is unknown or mysterious. One-down people attribute power to the one-up person and then attempt to psych that person out by deploying their antennas. The more that person is poker-faced and noncommunicative, the more people around that person will try to pick up clues—at least at the beginning, for if no information is forthcoming and that mysterious person shows no formal power and authority, others soon get bored of the game and quit. The mysterious person then remains in isolation. However, it is a possible ploy for a short while.

To be "unknown" is a power ploy. To be somewhat aloof, cool, distant, formal is a way of intriguing others and challenging them to get closer. When I was younger and dating, playing hard-to-get was the game. I am not advocating playing hard-to-get-known; I am saying that this strategy is either consciously or unconsciously used by many powerful people. Don't fall victim to it, but have it up your sleeve as an alternative if you need it. I must admit that this is something I have never been able to do. I like to be known (as you may have noticed) and will at times disclose more than

The Dynamics of One-Up One-Down Relationships

Feels good, free, satisfied
in the relationship

Experiences no need
to share feelings,
opinions, and judgments

Needs are largely met
by one-down person

Implicitly, keeps others
doing all the psyching out

Person who is one-up

Person who is one-down

Withholds feelings and
thoughts because of
perceived high risk

Spends a lot of energy
psyching out the
one-up person

Seeks out
support
from same
race and sex

Takes care of perceived
needs of one-up person,
at least enough to survive
in the relationship

Becomes resentful, frustrated,
and angry

Realizes own needs are
not going to be met

May lose track of
own needs

*Stan Hinckley, June 1977. Reprinted by permission

necessary and then regret it. Better to say too little in unknown or high-risk situations! Of course, friendships and good collegial relationships are based on mutually predictable behavior, which means sharing feelings and becoming known. There is a difference, however, between disclosing a feeling or thought and making revelations about your personal life. The former is a good role model we women can take; the latter may be inappropriate in a work situation.

The dynamics of one-up one-down relationships permeate most male-female interactions and are encountered on a daily basis. The entry of professional women into the business world as equal partners with men makes obsolete some of the traditional patterns of relating between the sexes. However, this entry is so recent that there are no precedents to follow. Without old models, we must create new models of our own.

NEW ATTITUDES AND BEHAVIORS

The men don't know anymore how to act. They say they are confused. Do they open doors for women? Do they light their cigarettes? Who picks up the check at a business luncheon? Who orders the wine? Who asks whom for dates? Who mixes the drinks? The guidelines are really not so difficult. They are: Whoever wishes to do it or whoever does it best!

Let us take some examples: Whoever happens to be in the lead should open the door and hold it for the other; whoever first sees a taxi should hail it. People emerge from an elevator in a logical procession, the front people first, the people in the back last. We should put on our own coats and lend a helping hand to a friend. Pick up a check in a coffee shop or a restaurant when it is your turn. A woman can invite a man for lunch and sign the check with a credit card. Dinner is seen as a more social occasion and it may be more difficult for a man to let a woman pay. She can either light her own cigarette, or let him light it and reciprocate the gesture by lighting his.[1]

When you go out drinking with the men in your office (which you should do occasionally) but don't want to drink, don't order tea or coffee. Order a soft drink (with a lime it looks better) and be sure to order a round of drinks when it is your turn to pay. Do not get drunk and indulge in nonprofessional behavior. If the men do it, they won't be talked about—but you will be.

Never sleep with your boss, *never* sleep with a client, *never* sleep with a married colleague. It is bound to create problems. You will be accused of taking unfair advantage of your sex. Besides, it is *always* the woman who either gets fired or stuck with the reputation, never the man. To invite a man up to your apartment or to go to a man's apartment is still seen today as an acceptance of a sexual tryst, so be clear on expectations.

If you are invited to a dinner party, you can go alone or ask the hostess if you may bring a friend. If you are the hostess, there is really no need to have sexes evenly matched for a dinner party. If your boss invites you for dinner, ask him if it is a business dinner or whether his wife will be joining you, and *act strictly professional* on either occasion. To reciprocate, you can ask them both to your home for dinner, or you may invite him alone for lunch. I am stressing all this because it takes so little for a man to jump to conclusions.

You can share problems of a personal nature with your boss only if you have a collegial relationship, but you should feel free to do so with your colleagues, if the sharing is reciprocal. There is a real double bind for women in terms of socializing with female secretaries. On one hand, they can be part of an important information network and support system; on the other hand, in order to be seen as promotable you have to socialize vertically (up) or you will be seen as "one of the girls."

To be seen with the "right" people once in a while is important, for other people will assume you are friendly with them, and knowing the right people is still one of the major sources of power and influence. You will have to continually juggle what you need, want, believe in, and must do to keep moving up. The attempt to keep your integrity while being political is an energy-draining task, yet that is the task of the coming decade.

If, in your organization, clients are taken out for lunch, then you should do so. But again, be careful of the misunderstanding it can create, for many men have never been taken out by a woman and may misinterpret the gesture. It is easier to ward men off *before* they make advances. Some men take women out and test them about their sexual availability just as a matter of course. If possible, they should be headed off before this happens. Joking will work better than anger. One way to avoid passes in restaurants when you are eating alone is to carry a brief case, which is the badge of the businesswoman.

At the office, if you are asked to get a cup of coffee at a meeting, ask one of the men to help you. If asked more than a couple of times, turn to one of the fellows and say, "Oh, would you mind this time?" Do this often so that the men begin to understand that it is *not* your role. Do not ask your secretary to get your coffee unless you sometimes reciprocate; however, if you are in the middle of a meeting and you need coffee for visitors, it is appropriate for you to ask your secretary to bring it in for everyone. If you have brought plants to the office, then you are responsible for them. If they are office plants, you can ask your secretary to be in charge of watering them. If you are expecting visitors and must tidy-up your office, do not ask your secretary to do it for you, do it together. Remember that your secretary is not a servant but a member of your team and perhaps one of your most important allies.

MANAGEMENT STYLE

Is there conflict between your own way of doing things and the prevalent style of the organization? As a rule, a manager is judged on results, not on style; however, if one of our objectives is to humanize organizations, then the first steps towards that goal are stylistic. Women generally differ stylistically from men in many of the things they do—reacting to events, responding to people, and thinking of themselves, for instance. However, female and male managers can achieve the same results without necessarily going about it in the same fashion.[2] The role, the task, and the resources as well as the existing company norms will dictate much of a manager's behavior. A

THE MANAGERIAL POLES
Assess your managerial style

	Always 1	Sometimes 2	3	Always 4	
I instruct hastily.	x				I go over every possible last detail.
I never criticize.		x			I notice every error and say so.
I never praise.			x		I frequently praise lavishly.
My people's needs come first.				x	The organization's needs come first.
I pass on responsibilities whenever I can.					I do everything myself.
I trust mainly my gut.					I trust mainly my head.
My employees' personal lives are not my business.					I am well informed about the personal lives of my employees.

	Always 1	Sometimes 2	3	Always 4	
I do not check on my employees.				x	I supervise very closely.
Employees' conflicts at work are best resolved by themselves.			x		I confront all conflicts between my employees.
I do not welcome suggestions from my employees.		x			I act on the suggestions of my employees.
I make all the decisions alone.	x				I consult others when I make decisions.
I never go to meetings.					I attend all meetings.
I grab at all opportunities as they arise.					I plan every step of the way.
I keep all information in my head.					Everything is written down in triplicate.
I am inaccessible to others.					I am available to others.

This should indicate your tendencies. Are you satisfied?

manager is also a norm setter and your managerial style will profoundly affect the people in the organization.

These stylistic differences are expressed in three ways:

Who I am
What I want from others
What I give to others

Who you are, what you want, and what you give will affect the work climate of your organization. Everything you do will make a difference. A manager is a person in authority and her words and actions are magnified by the power of her position. Awareness of your managerial style is of critical importance because of the impact it produces on others. Although most people fall somewhere between the extremes, it is important to know toward which of the poles you lean.

Who I am is a cluster of personality characteristics. If you are a warm, friendly woman who prefers informality and collegial relationships and wishes to be known as a person, you will produce a very different climate than a woman who is formal and aloof. Be aware of your characteristics.

What I want is your expectations. Whether you want your people to be autonomous or to wait for your directions, to be meticulous or creative, will make a difference. Know your demands and your tolerance levels.

What I give is your reward system. Do you praise good performance or notice only poor work? Do you share your power by having others participate in decision making, or do you decide everything alone? Do you supervise closely and exert a great deal of control or do you trust others? Do you give frequent feedback—both positive and negative—so that your standards are known? Do you help your subordinates move up? Are you generally supportive?

Know your managerial style. Your personality, your expectations, and your reward system will affect your staff's motivation, satisfaction, and productivity. *Who you are, what you want, and what you give will directly influence what you get,* both in terms of output and in terms of relationships.

YOU AND YOUR BOSS

Your relationship with your boss is probably the most critical at work because of the power he or she holds over you. You are dependent on that person's evaluation of you as a manager in order to be able to use your skill to the fullest measure and to gain satisfaction from your work. Your boss will also recommend you for promotions. As a middle manager you should be having frequent and fruitful meetings with your boss. The purpose of these meetings is two-fold: (1) to gain as much information as you can about the needs of the organization, since the implicit expectations stemming from these needs result in the output of your department; (2) to give your boss necessary information about what is going on in your department. Meetings also offer a mutual problem-solving opportunity: You need not confront difficult situations alone—together, you and your boss make appropriate decisions.

However, you will find many places where the performance review, done in a perfunctory manner, a couple of hours once a year, is the one and only contact you have. This is not enough. If at all possible, request more frequent meetings in order to ask if you are indeed meeting objectives, and seek possibilities for improvement. Many bosses are "too busy" to see their employees. If that is the case, try asking your boss out for lunch, saying that you just want to talk because you need to know standards of performance and would like to chat about how things are going. This is not only legitimate, it also shows your concern beyond the immediate task to be done; it shows an attempt to understand the thinking of upper management.

We as women can sometimes have a privileged relationship with a boss, with whom other men do not share what is seen as high-risk information. Because of our particularly well-developed verbal communication skills, we are sometimes able to more easily inform a boss of what is going on in the department, in the organization, and among the people. I don't mean gossip, I mean information. Our special competence at communications should help us to open doors for others as well as ourselves in ways that have not been usual or frequent in many businesses. For example, you can be a friend to your boss and he or she can be a friend to you. It is a reciprocal relationship of great mutual benefit. However, you must always remember the hierarchy: Never go to your boss's boss with requests or complaints unless the situation with your own boss has deteriorated so much that communication is not possible. In that case you may ask your boss to go with you to the person higher up. Always state that you know the impropriety of the step, but that you are driven to it because of the particular circumstances.

Requesting feedback

Far too many bosses know how to criticize but don't know how to praise, which means that much of the work you do goes unnoticed. This doesn't feel good, but there is something you can do about it. You can *request* performance feedback from your boss. Say that you want to know the things you don't do well enough yet, but you also need to know what you have done well, for we learn best from our successes. The most difficult part of this kind of session is not to become defensive when criticized, but to listen to the criticism in terms of the other person's perception of you as a worker as well as of your work. It is important to ask questions related to *the data* upon which the perception is based, since unless you understand that data, you can not determine whether the perception is correct. After all, the perception may be based on intuition or hunch, or you may perhaps be unaware of the way you influence your boss's perception of you. It is absolutely legitimate for you to ask for behavioral data. You need to get that feedback and you need to push for it again and again—especially since many men are reluctant to give feedback to women.

The issue around giving negative feedback to women is that many men are afraid that women will respond stereotypically, by crying or getting very upset. Men see themselves as protectors; therefore they "protect" the woman for her own good and do not tell her what she should know to improve her performance. Very often, it is already too far down the line and she has really botched something before a man will finally say anything to her. Of course, as it is too late, she will become appropriately angry and upset, which will naturally confirm the stereotype that "women are too emotional." If feedback were an on-going process, it would never get to that point! Ask for feedback in order to improve your performance, but do *not* ask for approval!

YOU AND YOUR COLLEAGUES

The two-person work relationship depends on the task to be performed. You and your colleague can be *mutually interdependent,* which means that neither could do the job alone. You both work on the same projects simultaneously and a continuous back-and-forth dialogue is needed to complete the task. You can be *serially interdependent,* which means that one of you must finish a piece of the task before the other one can take it up. The third type of relationship exhibits no evidence of personal connection, but the two people work side by side (figuratively speaking) since they have the same overall production objective.

Mutual interdependence requires good problem-solving ability and a capacity to tolerate each other's working styles. One person could work slowly, making the other very impatient; one could be a quick, flashy, creative thinker, whereas the other's style is to dig deep in order to understand the ramifications of each point. In order to get along with the other person, you both need to keep processing your styles, tolerances, and expectations of each other. For instance, if the person you are working with does something irritating, it would be more fruitful for you to say, "When you jump about from topic to topic it irritates me because I lose the continuity of my thoughts," than to say; "Why don't you stick to one thing at a time?!"

Mutual interdependence is a collaborative mode, but it contains seeds of competition. If each person is unsure of the other's opinion and wants to "look good," it is often difficult to focus on what needs to be done because you are focused on how competent you will appear. The newer you are and the less feedback you have received, the more tentative you will be, and therefore the more anxious. One way to avoid this in a two-person relationship, especially for women who may be newer at this than a male colleague, is to process the modes of interaction soon after you have had a chance to experience them. For instance, after having worked together for several weeks, I would say, "It is time to take a look at how we are working together. I, for one, am satisfied with this and that, but feel we could improve on such and such. How do you feel?" or, "I really like the way you handle this, but I wish you could do more of that," or "I'd like you to stop doing this other."

This opens the door to true colleagueship, something men have with each other but do not as yet have with women. Somehow for men, sharing feelings with women is connected with intimacy and, therefore, with sex. The trick here is to achieve *professional* intimacy, which means having a good understanding of each other and sharing task-related thoughts and feelings, rather than *social* intimacy, which is reserved for friends and lovers. For instance, be sure to send a note to congratulate a colleague who has achieved something. You will begin to form a network of goodwill.[3]

If you are serially interdependent with a co-worker, the main issue is expectations in terms of quality and deadlines. If you must produce before someone else can continue with the task, be sure that you and your colleague agree on the specifications and the date by which completion of your part is expected. If you sense that there might be a delay, be sure to warn the person who depends on you. Being forewarned can help someone be prepared. It is better to be upset when there is still time to deal with a delay than to be surprised with a totally unexpected lateness. If you are dependent on someone else's task completion, the issue may be the difficulty you have in voicing your dissatisfaction with either the quality or the timing. It is not easy to tell a colleague that the work is not up to your standards, or that you are angry because you were forced to wait. Being very clear about expectations is the first step. The second step is voicing your concern and dissatisfaction in a way that does not produce anger in the other person, but provides an arena for problem solving. For instance, instead of saying, "This won't do," you could attempt to talk about what is wrong with the other person's output. Sit down quietly together to see in what way there can be a change or an improvement; show a willingness to work on the problem. To give negative feedback and criticism is often difficult for women: It is uncomfortably close to rejecting. We fear to make the other person feel bad. One way to ease the pain of criticism is to tell the other person what they do well that you appreciate. However, there is nothing wrong in "straight talk." Eventually you will need to learn to do it without feeling uncomfortable or destructive.

One of the really difficult things for women to do is to disconfirm the stereotypes. People tend to hear and see what they expect to hear and see. When a person is exposed to behavior that does not fit a preconceived notion about how one ought to behave, the tendency is to disregard the evidence and continue to respond according to the stereotype. For instance, no matter how competent, aware, knowledgable, and skilled in giving positive and constructive negative feedback, you may still be perceived by one male subordinate as flirtatious, by another as competitive, by a third as manipulative, by a fourth as "too soft," by a fifth as a bitch. In other words, you will be seen as *whatever stereotype* each person has for female managers. But take heart. After a while, as you set a pattern of competent professional behavior, stereotypes *can* be reversed!

The onus on us women, of course, is that we are breaking new ground. Frequently, we are the first to attempt to break down prejudices and to help form new expectations of women as managers and as colleagues. Don't forget that most men have only known women as subordinates. We must keep repeating the appropriate behavior until it is part not only of our repertoire, but of theirs, too. A colleague of mine said that a behavior has to be repeated *seventeen* times before it becomes believable to others! My own experience verifies that this is not far from the truth.

Although more women are now moving into management positions, not all are welcomed when they get there. Many men still feel "invaded" in their male enclave. And so they are. Have you ever been part of a closed group, a club, a society, or a cluster of friends, where no stranger would be welcomed? Closed groups are very reassuring to the members because the boundaries are clearly defined. The people who are in feel good to be in and to have each other. By permitting outsiders, however, the inside group becomes less defined and provides less security and identity to its members.

When women come into formally male-dominated groups and organizations, the men feel the same way—somehow less secure and threatened with the fear of losing their identity. Even though none of this is conscious (so that pointing this out will not help), it still exists, and we women must be aware of this phenomenon in order to not push too quickly for inclusion. Just keep widening the crack a little more each day with your continuous determination for a collegial relationship, expressed in mutual help, information sharing, and fun. Yes, *fun*. Fun is a critical component of peer relationships at work. Socialization meets important needs; for many people, it is a primary reason for enjoying work. If you wish to maximize your effectiveness and satisfaction, socialize as often as you can "up," socialize frequently "horizontally," and once in a while socialize "down."

YOU AND YOUR SUBORDINATES

A good boss should know her subordinates well. What are their strengths, their flat sides, their preferences, their pet peeves? How do they work under pressure and collaborate as team members? How do they deal with ambiguity or unpredictability?

The Three Levels of Exchange
Between Speaker and Listener*

Feelings about subject matter

A   Subject matter   B

Feelings about one another

*Adapted from Cohen, *et al; Effective Behavior in Organizations*. (Irwin, 1976).

Are they meticulous plodders? Are they sloppy but creative? Are they responsible, autonomous, loyal, ambitious? If you don't know your staff, you can't utilize their strengths and help them develop their potential.

There is only one way to get to know people: spend time with them. Work and talk with them, then observe and reflect.

There are two critical components in the relationship with subordinates. First, a manager needs to *communicate* clearly with them; second, she must know how and when to *delegate* to them.

Communication

I like the drawing, left, because it clearly shows what goes on when any two people attempt to communicate. While person A communicates about a specific subject to person B, B not only hears what A says but also has feelings about the particular subject matter *and* about person A. These feelings will *color* what B hears. On the other hand, the way A speaks of the subject matter is colored as well by *A's* feelings about the topic and about person B.

For instance, if I say to a subordinate whom I really like, "Could you take care of this problem?" and the problem is no big issue for me, I would probably say it in a very calm voice, and my subordinate would most likely feel quite free to say he or she is too busy right now. However, if I have strong negative feelings about my subordinate or am very anxious about the problem that needs to be solved, then my message might be heard as an imposition or as a test of the person's ability to solve the problem.

The content of the message in both instances was the same. What got into the way was feelings about the message and about the other person. The only way to get around these barriers to clear communications is to keep checking with the other person on what they are hearing, on how they are feeling about the content of the communication. This is called *processing.*

The content is *what* is being said; the process is *how* it is being said. Women are very good at process observation. We see things going on that many men are unaware of. We pick up discomforts, anxieties, fears, and angers and often respond to them, not always sure of what it is we are responding to. Because we see so much, we tend to be helpful, because when we see distress in others, we want to alleviate it. Similarly, when we are upset, hurt, fearful, or angry, we expect men to pick up our cues and respond with understanding. When they do not, we are surprised and label them "unfeeling," not realizing that they simply did not perceive what we, in their place, would have seen. What this means is that we women must say much more about the way we feel than we really would like to. We must explain ourselves more than feels comfortable. Men need to be educated to become better process observers. Let us teach patiently, for there will be resistances to overcome. But the final rewards are great: People who can hear, understand, and respect each other without necessarily agreeing—who can collaborate on teaching each other about what there is to learn, both *out there,* being competent professionals, and *inside ourselves,* being aware persons.

It is only through this process that women and men will become known to each other without being threatened, without ever putting each other down. Instead, we can appreciate the similarities as well as the differences and celebrate both.

Giving feedback

Feedback is *the* tool for improved communication. The most important step for a manager to take before giving feedback is to gather information. However, in order to gather good information *you have to be open* to both good news and bad—even about yourself.

You will only be seen as open to information—especially if it pertains to your own managerial style—if you listen without becoming defensive. Do not try to explain why you did that particular thing that particular way on that particular day; rather, listen to what the person has to say, find out on what factors the data are based, and discuss the person's feelings about that information. You can then share how you feel about what you just heard. But it is *not* possible to say, "What you perceive is not so." You cannot disclaim another person's perception, just as no one can disclaim yours. You *can*, however, attempt to correct any misleading data, which may correct that perception. If someone says to you, "I see you as too demanding," answering, "No, I'm not too demanding," will not alter the person's feelings. You need to find out where exactly that person feels too much pressure and in what way the two of you can attempt to solve the problem. And don't take every piece of criticism to heart—remember, *one* action alone does not represent you (or anyone); only a discernible pattern does.

What we have just been talking about is feedback as you *receive* it. Now let us take a look at feedback as you *give* it.

1. *Feedback should be descriptive rather than evaluative.* In other words, describe the particular behavior but do not pass judgment on it. For instance, rather than saying, "You are too slow," say, "When you did this particular piece of work you did not meet the deadlines. What can we do about it?" or "What do you think is preventing you from meeting deadlines?" You may also share a feeling, if it is appropriate, such as, "It upsets me, because I must then change my plans." Then see what can be done about it. When this dialogue is finished, ask the person whether they agree with what has been discussed.

2. *Feedback should be specific rather than vague.* Do not say, "You seem so irritable today"; instead, say, "When I asked you about this project you shouted, and I don't understand why. It made me wonder if something is wrong."

3. *Feedback can deal only with behavior the person can do something about.* Do not say that the person is compulsive, or rigid, or a procrastinator. Accusing someone of a personality trait is not going to help. It will only make the person feel badly. Try to identify very small behaviors that can be changed.

4. *Feedback must be given when you are alone and have time to talk about it.* It is not appropriate to give feedback as you are leaving, or when others are present, or in noisy surroundings. It should also follow the event you are discussing as closely as possible.

5. *Feedback must be understood by the recipient.* Always check to see if the person can relate to your perception, allowing for disagreement. Suggest whenever feasible that other people's perceptions be obtained, since yours may not be correct.

The *trap* of giving feedback is that we sometimes give it in order to fulfill our own needs either to punish or control the other person, to show how smart or perceptive we are, or to give advice. Before giving feedback, always ask yourself if it will really be helpful to the other person. How will it be heard? How will the person use it? Are you doing it for yourself—to vent your feelings—or are you genuinely trying to help the other person?

When you receive feedback, be *aware* of your own feelings about it because this is data for you. Get as much information as you need to make the feedback helpful. In other words, apply the same rules to yourself. Be sure that the feedback describes a specific behavior, that the feelings of both people are acknowledged, and that you have heard correctly. The person giving feedback tells a lot about herself. If the truth does not apply to the recipient, it applies to the giver. *Feedback should never be discounted!* I also see feedback as a gift, for most of us live in ignorance of how others perceive us. If you would appreciate someone saying to you, ''I see you doing such-and-such really well,'' or ''I see you doing such-and-such poorly and would like to help,'' then it is safe to assume that your subordinates would like to hear the same from you.

It is only through constant feedback in an atmosphere of *mutual* trust that people can improve on their performance and feel they are working in a climate conducive to growth and development.

When giving feedback to male subordinates, remember that men have been raised by women, and that part of their maturing process has been to detach themselves from female authority figures (mothers, older sisters, elementary school teachers). To be a ''real man'' is to *not need* nurturing and advice from women, but to be self-sufficient and tough. Yet here they are again with a woman in authority, telling them what to do and wanting to be helpful. Initially, being too helpful, giving out advice, and controlling too closely by checking on everything may be seen by many men as too motherly, and will evoke old responses. A collaborative mode, that of a teamleader, would be the most appropriate stance for a woman manager. Only when your credibility has been established can you function with a broader range, give advice if you feel it is appropriate, and even offer help when you deem it necessary.

Delegation

How do you *delegate* to men, as opposed to women? Is there a difference? Yes. Women lack experience in asking men to perform tasks—and especially to perform to expectations as to quality and deadlines. Because we have been used to being delegated to *by* men we have difficulty in delegating *to* men. Delegating to women has different constraints. We may have newly acquired feelings of sisterhood and wish to feel equal; therefore, we balk at the hierarchical system. We also need the support system available at work, which may be found primarily among our female subordinates, so we do not want to change those relationships by making demands and evaluations. If we understand our different reluctances to delegate to both men and women we can then begin to transcend them, for delegation is perhaps one of the most effective tools in the hands of a manager.

Remember that delegation is a *granting of authority, not an assigning of task*. It requires trust on your part. There are three basic types of delegation. In the first, the subordinates have minimal responsibility and must report for your evaluation on a fairly continual basis. In the second, the subordinates make most of the decisions but need your approval to continue to each succeeding step. In the third, your subordinates are wholly on their own and report results only at the completion of the task.

Here are some guidelines for delegation.

1. Delegate by the results you expect. You may explain how the task should be done if your subordinates request this information. Otherwise, allow them to figure out the best methods to accomplish the task.

2. Be sure that you mutually agree on performance standards beforehand.

3. Agree also on checkpoints, so that you are not seen as distrustful of the person, but as fulfilling a contract to evaluate results along the way.

4. Be sure your subordinates have all the information they need. Make it a two-way process, so the subordinates can use you as a resource to gather more information as they go.

5. Give your subordinates opportunities to exercise freedom of choice and to learn from their own failures and successes.

What should you delegate? As much as possible. Delegate all daily routine work—in other words, work that is repetitive, because then it is worth training someone to do it. As the manager, you should concentrate on the things that need to be done for the first time, or on the things that need to be done only once in a great while. You are making better use of your time. Make a list of all your various activities and then divide them into three categories: (a) those things that only you can do; (b) those things that someone else in your department could start doing right away, or with very little training; and (c) those things that you can delegate, but require more extensive training. Next to that list write down the names of the people who could be delegated a task rather quickly, and then of those who need to be trained to perform the necessary work.

A good manager is always training her subordinates to replace her. Many women fear this, and perhaps justly; the room at the top is indeed very small. And the fear is that if you don't make it to the top you'll have nowhere to go and will just be replaced by someone you've trained. It is a legitimate fear, but to act on it is self-defeating. Delegation is not abdication. You remain accountable for the results in your department when you report to your boss. Let your subordinates be accountable for their performance if you give them the responsibility. Besides, subordinates appreciate most a manager who trusts them, gives them responsibilities, and allows them to take initiative and make decisions. You will head a team that has everyone pulling strongly in the same direction.

NOTES

1. Adapted from Letitia Baldridge, "America's New Manners," *Time,* November 27, 1978.
2. Anne Harlan and Carol Weiss, "Moving Up, Women in Managerial Careers," ongoing research (Wellesley, MA: Wellesley College Center for Research on Women).
3. Eliza G. C. Collins, "ABC of Executivity," *Cosmopolitan,* September 2, 1977.

SELECTED READINGS

Argyris, Chris. *Interpersonal Competence and Organizational Effectiveness.* Homewood, IL: Richard D. Irwin/Dorsey Press, 1962.

Athos, Anthony, and Gabarro, John. *Interpersonal Behavior.* Englewood Cliffs, NJ: Prentice-Hall, 1978.

Filley, Allan C. *Interpersonal Conflict Resolution.* Glenview, IL: Scott, Foresman, 1975.

Hall, Edward T. *The Silent Language.* New York: Fawcett Publications, 1959.

Johnson, Michael L. "Women: Born to Manage." *Industry Week,* August 4, 1975.

Rosen, B., and Jerdee, T. H. "Sex Stereotyping in the Executive Suite." *Harvard Business Review,* March–April 1974.

Rowe, Mary P. "The Saturn's Rings Phenomenon: The Progress of Women in Educational Institutions." Unpublished paper. Cambridge, MA: Massachusetts Institute of Technology, 1973.

Walton, Richard. *Interpersonal Peacemaking: Confrontations and Third-Party Consultation.* Reading, MA: Addison-Wesley, 1969.

Wells, Theodora. *Keeping Your Cool Under Fire: Communicating Non-Defensively.* New York: McGraw-Hill, 1979.

Chapter 10

DID I SOUND OK?

Fifty people at the meeting—
I want to say something,
But is it relevant?
And is it pertinent?
And is this the time
Or should I wait?
Perhaps it's dumb
Or has been said?
I wish it were not so important
for me to sound clever
and original
whenever I talk,
wanting every time
to make an important contribution
to the goings on.
I hate wanting others
to respect me.
I hate caring so much
that I should be liked.
Why should it matter—
But it does,
Damn it, it *does*!
So with pounding heart
I say it . . . .

Was it okay?
Tell me—how did I sound?

Women in Groups

Planning committees, task groups, project teams, executive meetings, departmental meetings, annual meetings, late meetings, productive meetings, endless meetings, boring waste-of-time meetings. . . . As a manager, you will attend them and you will call them. Either way, whether you are a member or in charge, your group skills can increase the effectiveness of your group.

You can have two formal roles in a group: participant or leader. As a group leader you are in charge. You can take a leadership role (for example, setting the agenda) or you can take a facilitative role (helping others set the agenda). As a participant you can be an informal leader, but it is a role that you take on yourself, not one that is assigned to you. You may also be a chairperson, who neither leads nor participates but help others do so; it is a facilitative role.

Before we take a look at women as both participants in and leaders of groups, it is important to understand in what way women and men behave differently when in groups. The following chart summarizes research on these differences.[1] I find it helpful just to recognize that some of my behavior patterns are very similar to those of other women. First, it is comforting to know that I am not the only one feeling this way; it makes me less hard on myself. Also, if I can recognize a problem and be aware of its accompanying behavior, both in me and in others, I have more control over what I do.

When men and women are in mixed groups, different behaviors emerge. Men begin to interact more with individuals, especially with women, but women direct their remarks to the more dominant men. Even though "women in all-female groups compete little with each other, women in male and female groups appear to spend a major part of their time in competition with each other for male attention and approval."[2]

When they get together in mixed groups, women talk less and feel more restricted, while men expand their repertoire by changing to a more personal orienta-

Men's behavior in all-male groups	Women's behavior in all-female groups
The same men tend to be the active participants for the life of the group, never missing any meetings.	Patterns of activity and dominance change from meeting to meeting.
Low participants often miss meetings or drop out entirely.	Women who were active at one meeting draw out the other women at subsequent meetings, saying they were uncomfortable if they were "too dominant."
Remarks made to the group as a whole, not to individuals. These remarks are impersonal and indirect.	Remarks addressed to individuals 90 percent of the time. These remarks are personal and direct.
Generalize—speak for others.	Personalize—speak for themselves.
Topics deal with competition, aggression, and superiority and are task-related.	Topics deal with self-revelations and expressions of feelings and are process-related.
Camaraderie achieved through discussion of events and joking.	Closeness achieved through discussion of self and family.
Identity and status based on achievement.	Identity and status based on relationships—how they perceive each other.
Jump from one topic to another.	Discuss same topic for half an hour or more.
Exhibit competition, an attempt to win, and a power orientation.	Exhibit collaboration, an attempt to please other group members, and a support orientation.
Show strength, hide weaknesses.	Exhibit weaknesses and show vulnerabilities; hide or repress strength.
Anger expressed with ease.	Anger expressed with difficulty.
Out of touch with own feelings; feelings repressed or denied.	In touch with own feelings.
Verbalize feelings with difficulty.	Verbalize feelings with ease.
Blame exteriorized: "It's your fault" or "It's their fault."	Blame interiorized: "It's my fault."

tion, showing greater self-disclosure and less competitive behavior. In a study of 1,500 women who were included in "Who's Who in American Women," it was found that a significant number graduated from women's colleges rather than from educational colleges and universities.[3] Since a group (the class) is the typical setting for learning in institutions of higher education, it is not surprising that in mixed groups or classes women would be dominated and therefore talk less, ask less, and learn less than they would in all-female groups. Female role models in the form of female instructors have also been shown to be an important element. If participants can identify with the leader, they can aspire to emulate her and to eventually become leaders themselves. Other data also suggest that women in all-female groups experience greater positive gains than women in mixed groups.[4] Men, on the other hand, do better in mixed groups in terms of learning and attitude changes.

What can we make of all this? Well, next time you are in a group, use this knowledge to plan your strategy. Talk to individuals *and* to the whole group. Be task oriented *and* pay attention to process. Be confronting *and* supportive. Generalize, but personalize when appropriate. Blame others when it is *their* fault—take the blame only when it is yours. If anger is the correct response, then show it. Cooperate when that is necessary, but compete when you want something for yourself. Show both your strengths *and* your weaknesses. In other words, be a whole person, showing all sides of yourself. By expanding your range of behaviors you will become richer yourself, and you will model to others the infinite possibilities of both *initiating and responding*. The most valuable member in the group is the androgynous person!

PHASES OF GROUP DEVELOPMENT

To become a skilled group member or leader you must understand group dynamics. What happens in a group will depend on many things, such as the nature of the task, the setting, and the makeup of its membership. But whatever these factors may be, groups seem to go through predictable developmental phases. If you know what to expect, you can better understand what is going on and why it is happening and can then act accordingly.

Think of children growing up. They will first be dependent on their parents; they will need their approval and will follow their directives on what to do and what not to do. As they grow up and become adolescents, they go through a period of counterdependence. We all know of the adolescent rebellion, when children challenge parental authority. They will not do as they are told (noncompliant). They want to become autonomous but do it awkwardly, in an exaggerated manner. After the adolescents develop into adults they begin to see their parents not just as figures of authority but also as people with whom they can relate on an adult level and whose approval is not always necessary. Young adults wish for total independence and autonomy.

As they become older, they can return to their parents as friends or colleagues and accept an interdependence, in which parental approval is not necessary but is pleasant. They are able to interact with their parents and with other figures of authority in a noncompetitive manner: they can both give and take. They no longer have to prove anything.

Phase One   A group that exists over time develops just like a person. A group starts out by being, like a child, *dependent* on its leader. The group members look to the leader to help them set goals and define objectives. If there is no clear leader in the group, weaker members will tend to be dependent on the stronger members. One of the central issues of a new group during its *forming* period is who is to be ''in'' or ''out''—who will be the good members, who will set norms, who will follow them, and who will be the deviants? The issue is one of *inclusion* and membership.

Phase Two   As the group progresses, it goes into a stage similar to adolescence, which is *counterdependence*. In this phase, group members vie for *control* over the group. They either overtly challenge or covertly sabotage the leader. Members either drag their feet, ''forget'' to perform tasks, or talk about the leader behind his or her back. This stage is also known as *storming*. Members who emerge as leaders are followed for a while, then scapegoated or seduced to be more participative.

Phase Three   As the group emerges from its power struggle, it goes into what is known as the *independent* or *norming,* phase, which means that the group is establishing norms by which to work together more effectively. There is no need for strong leadership. Members are used as resources for the task. Group members feel *affection* for each other and like to work together.

Phase Four   The final phase is *interdependence,* also known as *performing.* Members work as a unit and can be both task- and people-concerned. Power and influence are distributed among those who have specific expertise and can change according to these measures. The leader is a participative member. There is no issue over power, control, or inclusion unless there is a deviant member, who is dealt with by either exclusion or confrontation. The group members feel an *intimacy* with each other and gain a great deal of satisfaction from the exchanges that are possible within the group.

I have included here a chart to help you visualize the sequence of growth a group goes through. As you read across the page, remember that all three words in a group refer to the same phase. This will give you a clearer image of what those phases are about.

THE WOMAN AS PARTICIPANT IN PHASE ONE

Phase one: inclusion/dependence/forming. There is already a special issue for women: inclusion. If you are a token woman in a group, or among a minority of women, there may be an attempt by the males in the group to exclude you and to form a male bonding group. If you see this happening, you need to *confront* it and

Phases of Group Development*

One	inclusion	dependence	forming
Two	control	counterdependence	storming
Three	affection	independence	norming
Four	intimacy	interdependence	performing

*Adapted from William C. Schutz, *Elements of Encounter* (Big Sur, CA: Joy Press, 1973); W. A. Bennis, "Patterns and Vicissitudes in T-group Development," in *T-group Theory and Laboratory Method*, ed. by P. Bradford, J. R. Gibb, and K. D. Benne (New York: Wiley, 1964); and B. W. Tuckman, "Developmental sequence in small groups," *Psychological Bulletin*, 1965.

claim membership. How do you do that? You participate. You take on task roles. There are five main roles that you should consider taking:

1. **Initiating,** which means proposing or suggesting ideas, actions, or procedures.

2. **Informing**—asking for facts, ideas, feelings, and opinions of the members or offering your own ideas and feelings to the group.

3. **Clarifying,** which means explaining and building on ideas or suggestions of others.

4. **Summarizing**—pulling together what has been said or done so far to help the group consider where it stands.

5. **Testing the consensus**, which means exploring whether the group is nearing a decision. This will either prevent premature decision making or help a bogged-down process.[5]

A work group is by definition task-oriented. A *male* work group is doubly task-oriented since men in groups are focused on tasks. Therefore, the initial inclusion phase is too early for women to deal with process questions. To focus on "how" instead of "what" will be seen as breaking the norm. An example of paying attention to task would be to say, "Let's decide on this now. All those in favor, raise their hands." An example of paying attention to process would be saying, "Let's discuss how we will make decisions. Is voting what we want to do?" To do the latter is to take on a leadership role, which may be what the group needs but may also delay your membership in the group. However, you can take that choice if your level of self-confidence is high enough.

In this early phase of group development the lone woman or the two or three in the minority will often be tested. This may take the form of joking, making sexual innuendos, or ridiculing or discounting whatever she says by flirting with her, calling her Mother, or patronizing her. What to do? If you say you dislike the joking, you'll be told you have no sense of humor; if you are upset at their discounting behavior, they'll say you are too sensitive; if you get angry, they'll complain that you're too emotional. In the face of opposition, I try to remain task-oriented and professional, to speak out loud and clear, to repeat what I've said if I'm not heard, and to seek eye contact. The longer you wait to speak up, the harder it becomes, so do it in the first few minutes of the group.

THE WOMAN AS LEADER IN PHASE ONE

As a leader, you know that in this phase the members will look to you for direction. This means setting the agenda, making decisions as to how to proceed, and prioritizing discussion topics. But as a woman "in charge" there is already potential trouble. In its dependent phase, the group may actually be more counterdependent because of the initial resistance that men may exhibit toward women leaders.

The nonacceptance of a female in authority may lead to competition for her position from the male members, and therefore to an inability of the group to work effectively or to a discounting of the woman in charge. If this happens at the beginning of the group, when control should normally not yet be the issue, it needs to be confronted. How? By taking a firmer stance as the person in charge, by asking the male who is competing, "Are you competing with me?" or, "What is your proposal, if you don't like mine?" Make a process comment, such as "What is going on here?"

As a leader in a group you are a norm setter. What roles can you take to help the group function productively?

1. **Harmonizing**—reconciling disagreements, releasing tension, helping people explore differences.

2. **Gate-keeping**—keeping communication channels open, suggesting better procedures, bringing silent members in by asking their opinions.

3. **Encouraging**—indicating with words and facial expressions that the contributions of others are acceptable, being warm and responsive.

4. **Compromising**—modifying your position if appropriate so the group can move ahead, admitting errors.

5. **Giving feedback**—telling others in helpful ways how their contribution is received or how their behavior is perceived.[6]

I hope you see the difference between leading and participating in the first phase of group development. As a group leader, you can be a norm setter; as a group member, you may want to be careful not to break norms, for it may be that having a woman in this group is already breaking a norm for the men present.

THE WOMAN AS PARTICIPANT IN PHASE TWO

Phase two: control/counterdependence/storming. You can expect this phase to emerge approximately 3/5 into the time of the life of the group. It does not seem to matter how long the group lasts; whether it lasts one week or one day, 3/5 into the group life control issues will usually emerge. If I teach a semester of fourteen weeks, I expect problems in my classes approximately during the ninth week of classes: Students will complain about assignments, about my style, about each other—whatever, they will complain. Or if they do not do it overtly, they will come unprepared, not come, come late, do poorly. I have learned a coping strategy: Approximately during that time, I open the class up for an evaluation, which I call the mid-semester evaluation. This allows the students to vent their feelings and to express their need to control the content of the material being presented or the teaching process.

If you are a group member in the second phase, you will see leadership being challenged, and will either compete actively for the leadership yourself, will help others compete for it, or will attempt to defuse the strong feelings being generated. This is a particularly difficult phase for women because it has not been seen as legitimate for women to compete for power. Time and again you may be discounted in your efforts and may also, time and again, be reduced to silence. Men have very subtle ways of doing that. One way is never hearing when a woman makes a contribution; another is attributing what she has said to someone else or calling it the "woman's point of view." This can make things very hard for you, but if discounting behavior is going on, you can point to that fact. Odds are that after such a challenge, the men in the group will listen to you. Of course, the onus is then on you to sound smart, or else they'll ask, "Why did you make all this fuss?" Either way, it's not easy.

For a woman to do all this, she needs to have an enormous amount of self-confidence. If you can't do it, and I certainly identify with the difficulty of that kind of risk taking, then there is still another thing you can do: You can talk to the members individually, outside of the group. Discuss your problem with them and ask if they can (or are willing to) collaborate with you in changing the group norm. All you want is some space for you to be able to take your share of the influence in the group.

THE WOMAN AS LEADER IN PHASE TWO

You are in charge and they are rebelling. If you know that this is a normal phase of group development, you will not tend to personalize the rebellion and think, "What am I doing wrong?" Instead, you will know that the group needs to get through this phase in order to get to the next one. Phase two cannot be by-passed, just as adolescence cannot be by-passed. The rebellion may take several forms. It can be an out-and-out challenge of your leadership, it can be a very subtle disagreement, or it can be noncommitment to the task or process you suggest. It can go anywhere from taking up a whole meeting or a series of meetings to just taking up a few minutes. You need to ride out this phase without feeling touched personally. The decision to be made by you at this time is how much do you trust the group members to take control, and how much control do you need to retain?

This goes back to your management style. What are your control needs? How much do you need to be in charge? How much can you trust others to take over? What are the needs of the group? Leadership does not need to be exerted uniformly. Different people can be in charge of different aspects of a task; as leader, you can assign such roles. For instance, you can ask someone to be in charge of time keeping, someone else to be in charge of taking minutes, another person to be at the blackboard or flip chart, someone else to give the report, and yet another to offer expertise. A word about newsprint or blackboard writing: I like to take this role because I feel that the person who writes down whatever the group produces has visibility and power. That person asks for clarification from the members but ultimately decides what is being recorded. It is not the same thing as the secretary who takes notes on a pad. So if ever given the chance, remember that the role has influence and *stand up there.* On the other hand, if you want to share the influence, you may ask someone else to do this.

THE WOMAN AS PARTICIPANT IN PHASE THREE

Phase three: affection*/independence/norming. As the group moves out of the storming phase toward independence, the group members attempt to articulate what some of the norms of the group are to be in terms of feedback, decision making, conflict management, and leadership. The leader of the group becomes more of a member, and the members have affection for each other. For the woman participant in a group at this third phase of development, the problem may be centered around affection. I do not necessarily mean overt expressions of affection but expressions of enjoying working together, of feeling independent of any leadership, and of being comfortable with each other and with the established norms. Men show affection toward each other by backslapping, rib shoving, and arm punching. This type of expression is not directed toward women, nor do women do it to men. Women tend to give and receive hugs, squeezes, pecks, and affectionate smiles.

If you see the men being physical with each other, be physical also in whichever way is comfortable. Physical touching breeds eventual comfort with each other and it is what you want to achieve in a group. I have a whole repertoire of nonsexual expressions of affection: shoulder squeezing, knee slapping, hair pulling, arm patting, shoulder bumping, gentle toe-kicking. It all seems to work quite nicely.

THE WOMAN AS LEADER IN PHASE THREE

Women should be able to relinquish leadership to other members. We also need to deal with behavior of members that interferes with the work of the group. Some of these self-oriented behaviors are:

1. **Aggressing**—attacking, deflating, using sarcasm.

2. **Blocking**—resisting beyond reason, using hidden agenda items that prevent group movement.

*William Schutz uses both affection and intimacy as the third phase of group development.

3. **Dominating**—interrupting, asserting authority, participating to the point of interfering with others' ''air time.''

4. **Avoiding**—preventing the group from facing controversy, staying *off* the subject to avoid commitment.

5. **Abandoning**—participating too little, being obviously not involved.[7]

These self-oriented behaviors need to be dealt with because even though some of them are appropriate for the second stage, they are dysfunctional as the group moves towards independence. Again, how to confront? If you can identify the behavior, you can challenge the member with this particular way of acting. Say, for instance, ''I understand your point of view, but would like to hear from the others also,'' or, ''You don't participate enough; I miss you as a lost resource,'' or, ''I don't think sarcasm helps here,'' or, ''What are you *really* resisting?'' or, ''We need to face the conflict, not avoid it.'' These are all process comments—they are not related to the task, they are related to how the group is going about its task, and they are legitimate comments from a leader to keep the group on target.

THE WOMAN AS PARTICIPANT IN PHASE FOUR

Phase four: intimacy/interdependence/performing Your group has arrived at the final stage when members can share not only professional but also personal concerns with each other. They are supportive, they show that they need each other's expertise, they work as a team, and they share leadership by taking on various needed roles, and then interchange them. This is a time when the group generates excitement and high energy. Some people can get a real ''high'' working in such a group.

Constructive feedback is a norm. People do not threaten each other. There is a climate of honesty and trust. If a group has truly achieved this fourth phase, there should be no issue for even a token woman, for she is accepted as a true member, without being typecast as mother, daughter, or sex object. The group is performing at a high level and she should find enormous satisfactions from working with such a group.

THE WOMAN AS LEADER IN PHASE FOUR

The leader of a group at this stage is more of a participant than a leader. Power not being an issue, she may only need to intervene in the case of a crisis.

The diagram of the three interlocking curves that follows represents the issue most in focus in the different stages of development. The first curve deals with whether you are going to be in or out. As this issue is resolved, the next one starts: Who is going to be in charge? Who is going to be active or passive? As this issue gets resolved, the next issue comes to the fore: Are the members of this group going to be close to each other or distant? Who will be close to whom and who will remain more formal? As you can see in the diagram, the issues interlock, but at various phases of a group's life, different issues take on more importance. This does not mean that all issues cannot be recycled. For instance, even toward the end of the life of a group, a member who has always felt excluded may voice either pain or anger around that exclusion. Or some fighting over leadership may occur either early or late in the group's life, and not when it is usually expected.

These phases, as I described them, are meant to be a help to you so that when they occur you feel that they are normal, to be expected. They are *not* immutable. Variations do happen, and at times you may think a particular group you are in does not seem to follow any pattern. However, as a rule, you can expect to see predictable sequential phases in most groups.

Knowing this, whether you are a participant or a leader, should help you understand behavior, yours and others! It is only when you *do* understand that you can deal with it effectively.

When a group must terminate after having gone through the four phases, there will usually be sadness and often an attempt to provide the group with new tasks in order to prove the necessity of its continued existence. As leader of such a group you should be wary of this phenomenon. If the group has indeed fulfilled its purpose, then the group leader should call a mutually agreed-upon termination time.

I believe that endings are made easier if they are finished with a celebration of some sort. In other words, a ritualistic termination makes it easier to accept the finiteness of the group. It can be a bash at someone's home or sherry at the end of the last meeting, in order to celebrate the work achieved.

Seating arrangements

Close attention must be paid to the seating arrangement of any group. Sit at the head of the table if you are the chairperson or the leader. It will confirm your formal role. If you are a participant, either sit next to the leader or opposite the leader, if you wish

Phases of Group Development*

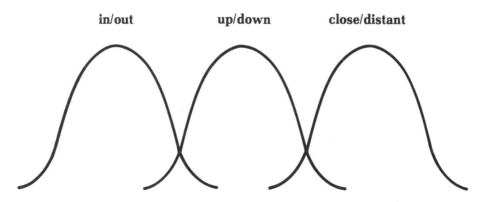

in/out **up/down** **close/distant**

*From William Schutz, *Elements of Encounter* (Joy Press, 1973).

to have any influence. Even though research demonstrates that people who sit at opposite ends of a table are usually antagonists, by sitting opposite the leader, you will not only catch his or her eye most often, but you will also be the person to whom most of the comments are addressed.

Never sit with no one on one side of you—in other words, in the end place. If you do, you will have half the possible interactions all the other members have. If there is a circle of chairs, take the most comfortable or largest one. If there is a window, sit with your back to it; you will get a good light on other people's faces. Do not always take the same seat unless it is the power seat, the seat seen as belonging to the leader. Groups whose members always sit in the same places relative to each other seem to be more static than groups who are able to interchange places. If you never change the seating, you will see only one-half of the two faces on either side of you most of the time, and by changing seats you will get a new perspective on those people.

Coming to meetings early assures me of my preferred seat. However, if I come too early and do not know where anyone else will sit, I may end up by not having the juxtaposition I wish. You may think of this as trivial, but it really isn't. Research on small groups has demonstrated that seating arrangements affect the outcome of a meeting.[8]

HOW DO YOU RUN A MEETING?

As group leader, it is your responsibility to control the climate and the direction of the meeting. A few points are essential to remember:

1. *Always* have an agenda. You either come prepared with one, or you set one with the members of the group. If you have an agenda of your own, explain your objectives and get input from others.

2. Decide right away at which time you will break up the meeting, and stick to that time. If controversies arise or people seem tired, take a five-minute break.

3. Decide who will be taking notes, if they are necessary. It should not be you, if you are in charge.

4. The role of chairperson is like that of an orchestra conductor: You recognize people who wish to speak but have relatively little or no input yourself.

5. You must exercise greater control when:
   a) the meeting is oriented toward the exchange of information.
   b) the topic generates strong, potentially disruptive feelings.
   c) the group is moving toward a decision.
   d) time pressures are significant.[9]

6. Group problem solving is encouraged when:

  a) no single individual is allowed to dominate.

  b) assumptions can be safely tested.

  c) alternatives can be explored.

  d) the focus is on valid data related to the problem.

7. Be sure your role is understood and accepted by the people in the group. If you feel the meeting is diverging from its task, then all you need to say is "I feel we are diverging from the task."

8. Never break up a meeting until everyone understands what needs to be done next and who will be responsible for it. A meeting that ends with a vague "We should be doing such-and-such next" is *not* a meeting that will produce anything concrete.

9. If another meeting is to take place, set the data while everyone is there and decide on a tentative agenda. If the meeting is not going to take place for a while, I like to send reminders to people, adding the agenda to my note.

There are three main problem individuals in any group: the person who talks too much, the person who talks too little, and the person who talks off-target.

I visualize a large pie sitting in the middle of any group of people. That is the available "air time." If one person takes a very large slice, it will leave very little for others. Every time a person takes a piece of the pie (or the "air time"), he or she ought to replace it with *input* for the group's task! So those who take very large chunks of pie ought to be giving very large chunks of input. The person who takes no pie does not have a chance to put anything in.

Why do some people talk too much? Well, they may have expertise which others do not have. They may want to be helpful; they may be overly enthusiastic or impatient with the group's progress. They may be competing for leadership, or they may be seeking approval for their cleverness or effort.

Whether they are anxious to "look good" or just think highly of themselves, your handling of the situation is the same. Acknowledge the person's contribution and then say, "We would like to hear from others, too." Another tactic is to ask the talker to summarize what he or she has been saying and then ask for others to speak.

What about the person who talks too little? Again there may be a variety of reasons for not taking a share of the pie. The person may be bored with the topic or unknowledgeable of it. People often do not talk for fear of appearing incompetent. This is, of course, the problem of the shy person. Some people are slow thinkers; by the time they have formulated what they want to say, the group has moved on to another topic, or someone else has said it first. You need to help the noncontributor. A silent person is a lost resource for your group. Frequently, women fall into that category. Why? Women who have had little experience in groups may not have enough self-confidence to value their contribution. Because of women's generally low self-esteem, they often do not know *what* they know or do not trust that what they know is valuable.

If women are in a minority, talking out gives them high visibility. If one man out of twelve says something foolish, it is not noticed. But if the only woman, or one out of two, says something foolish, it is very noticed and remembered. Also women seem to compete less in groups, and that includes competing for air time. How can the group leader help the shy persons in the group? One way is to ask them what their opinions are, but it may be putting that person on the spot. One of the things I have done in the past is ask the chronic nonspeaker outside of the group if they would like some help by being called on. If the answer is no, I respect it. Another way is to ask some question of every member in the group, so that the shy person's voice is heard also. The longer the nonspeaker remains silent, the harder it is to break in, because invisibility increases with the length of time one is silent. A token woman in the group who is not used to putting in her "two-cents worth" and waits to speak only increases the difficulty of ever overcoming her block. As a leader you must pay attention to this problem early in the life of the group. If the whole group is silent, it may be caused by boredom, by fatigue, or by resistance to being there. People also will not talk if they sense a risk or sense that their contribution will not be valued. The way to deal with this is to make a process comment, say, "The group is very quiet. Are there issues that need to be dealt with that are not in the open?"

The person who makes task-unrelated comments may be sabotaging the objectives of the group either because the objectives are not understood or because that person disagrees with them. Perhaps the topic is too emotionally laden. The group leader should be careful not to punish this person. Instead, restate the objectives of the meeting, and if the person continues to sidetrack, then ask how what he or she is saying relates to the topic.

If there are side conversations between two or more of the members that interfere with the group's work, then you may ask if these people are saying something the whole group ought to hear or why they dropped out. If it does not happen too often, I would let it go because it just may be that someone did not understand something and is checking it with another person near by. But it may also be that that particular conversation is more interesting than what is going on in the group.[10]

The following chart will help you identify (1) the group issues, (2) the questions of group leader should ask of herself about the group, and (3) the individual members' dilemmas that each member needs to answer for themselves. If you work in groups, look frequently at this chart. It will give you a handle on group dynamics. And as you know, anything that helps you understand your work environment better will help you deal with it that much more effectively.

It is not easy for women to gain group membership, achieve inclusion, and become "one of them." If women want to achieve leadership in groups and have influence in decisions, we will have to keep pushing our way against all odds; keep asserting our rights and our power; keep breaking norms, stereotypes, and expectations.

ISSUES FACING A WORK GROUP*

Group Issues	Participant Issues	Group Leader's Questions
1. Atmosphere and relationship	How friendly and close do I want to be? Will others allow that?	What kinds of relationships should there be among members? How close and friendly, formal or informal?
2. Member participation	How much do I want to participate? Can I be as quiet or active as I'd like?	How much participation should be required of members? Some more than others? All equally? Are some members more needed than others?
3. Goal understanding and acceptance	Are the group's goals compatible with mine? If I have different goals will there be a place for me? Should I seek a more compatible group or try to change my goals?	How much do members need to accept or be committed to group goals? How much do they need to understand the goals? Everyone equally? Some more than others?
4. Listening and information sharing	Will I be able to get the information I need to do my work? To whom do I have to listen? Can I get others to listen to me?	How is information to be shared? Who needs to know what? Who should listen most to whom?
5. Handling disagreements and conflict	How freely can I disagree? Will others disagree with me? Can I fight for what I believe in?	How should disagreements or conflicts be handled? To what extent should they be resolved? Brushed aside? Handled by dictate?

Group Issues	Participant Issues	Group Leader's Questions
6. Decision making	Will I have a say in important decisions?	How should decisions be made? Consensus? Voting? One-man rule? Secret ballot?
7. Criticism of member performance	Will I be able to tell others how they are doing? Who will tell me how I'm doing?	How is evaluation to be managed? Everyone appraises everyone else? A few take the responsibility? Is it to be avoided?
8. Expressing feelings	How open can I be about my feelings?	How should feelings be expressed? Only about the task? Openly and directly?
9. Division of labor	Who will decide what I do? How much say can I have over that?	How are task assignments to be made? Voluntarily? By discussion? By leaders?
10. Leadership	Will I be able to exert influence on the group? What happens if I try?	How should leadership functions be exercised? Shared? Elected? Appointed from outside?
11. Attention to process	If I don't like the way we are doing things, can I say so?	How should the group monitor and improve its own process? On-going feedback from members? Formal procedures? Avoiding direct discussion?

*Adapted from Allan Cohen, Stephen Fink, Herman Gadon, and Robin Willits, *Effective Behavior in Organizations* (Homewood, IL: Richard D. Irwin, 1976). Charts based on the work of McGregor.

Women are not out to conquer; we merely want to find a good place for ourselves. For this, we will have to teach our-selves and others new behaviors and new attitudes:

Pioneers needed:
    terrain uncharted
    environment frequently hostile
    sustenance meager or nonexistent
    climate adverse
    results uncertain
    end of journey not in sight

To facilitate the inclusion of women in groups, there should be two types of opportunities available: One is for women to be with each other for support; the other is to be in mixed groups. The more opportunity for women and men to interact, the more they will become familiar with each other and the more chance there will be of bridging differences.

Groups that are tightly knit, with closed boundaries, will make it very hard for women to get in. However, once in, many women experience such a relief after the struggle that they do not move on to other groups, even though it might be wise for their careers. Who wants to struggle all over again continuously? The comfort of hard-earned affiliation or relationships is difficult to give up. This is an important concept to remember. Women need not only help and support in getting into groups, they need help and support to get out of them when the time has come to move on.

NOTES

1. Aries, "Male-Female Interpersonal Style in All-Male, All-Female and Mixed Groups," in *Beyond Sex Roles,* Alice G. Sargent (St. Paul, MN: West Publishing, 1977). Barbara Bunker, "Group Behavior and Women," in *The International Encyclopedia of Psychiatry, Psychology, Psychoanalysis, and Neurology,* ed. by B. B. Wolman (New York: Van Nostrand Reinhold, 1977). Rosabeth Moss Kanter, "Women in Organizations: Change Agent Skills," in *New Technology in Organization Development* (Arlington, VA: NTL Institute, 1974).

2. Charlene J. Carlock and Patricia Yancey Martin, "Sex Composition and the Intensive Group Experience," *Social Work* 22–1(January 1977).

3. Elizabeth M. Tidball, Perspective on Academic Women and Affirmative Action," *Educational Record,* Spring 1973, pp. 130–135.

4. Carlock and Martin, "Sex Composition. . . ,"

5. Adapted from K. D. Benne and P. Sheats, "Functional Roles of Group Members," *Journal of Social Issues* 4–2(Spring 1948):42–27.

6. *Ibid.*

7. *Ibid.*

8. Jean Brown Edson, "How to Survive on a Committee," *Social Work* 22–3(May 1977).

9. Adapted from James Ware, Case Clearing House, Harvard Business School. Copyright © 1977, by the President and Fellows of Harvard College.

10. Adapted from John S. Randall, "MO," *Training and Development Journal,* November 1978.

SELECTED READINGS

Bennis, W., and Shepard, H. "A Theory of Group Development." *Human Relations* 9(1956):415–437.

Bunker, Barbara Benedict. "Group Behavior and Women." In *The International Encyclopedia of Psychiatry, Psychology, Psychoanalysis, and Neurology,* ed. by B. B. Wolman. New York: Van Nostrand Reinhold, 1977.

Egan, L. *Face to Face.* New York: Brooks/Cole, 1973.

Napier, R. W., and Gersinfeld, M. K. *Groups: Theory and Experience.* Boston: Houghton Mifflin, 1973.

Schutz, William C. *Elements of Encounter.* Big Sur, CA: Joy Press, 1973.

Wolman, Carol, and Frank, Hal. "The Solo Woman in a Professional Peer Group." *American Journal of Orthopsychiatry* 45(1971):164.

Chapter 11

LEADERSHIP

If the best of me can make more of you
Then the best of you will reflect on me

Women as Leaders

Chief executive officer (CEO), chairperson of the board, president, director, senior administrator, head honcho, top executive. Since less than 1 percent of the women in the work force today are in these positions, it is not just to this population that this chapter is addressed. Rather, I am addressing the women who want to swell the ranks of that 1 percent. If we simply doubled the percentage every couple of years, in one decade we would hold one-third of all top managerial positions. Alas, geometric progression is not the way society changes.

However, women's progress in top executive positions is remarkable when compared to a few short years ago and abysmal when compared to men. F. W. Way, a twenty-year-old professional organization of 350 women working in banking, investment, and related fields, reported that at the end of 1978 there were 268 women serving as directors of Fortune's top 1,300 companies.[1] Of these, only 10 were earning more than $40,000. This is less than 1 percent, and the number is *not* growing. The ratio of male to female corporate chief executives stands at 600 to 1. Only a little more than 2 percent of the women in our present work force earn over $25,000 a year.

Why is there only 1 percent in top management, when women comprise half the labor force? Twenty-one female managers, whose careers were tracked in the Boston area from 1963 to 1973, were found to lack "goal obsession;" that is, they were not compulsive or especially shrewd in mapping out their career strategies. They did not have an overriding ambition to rise higher. It was suggested that this is due to the socialization process girls go through, which reduces their motivation, their ambition, and their achievement needs.[2]

That much is due to our socialization is surely true, but it is also true, as Jean Baker Miller writes, that society undervalues women's attributes and therefore underrates women's contributions.[3] Rosabeth Moss Kanter finds that the structure of the work environment does not offer opportunities to women and thereby prevents their advancement. Kanter's position is that *all* people—men *and* women—in low opportunity situations react the same. For instance, she believes that women do not have more affiliation needs than men, but that all people with no opportunity for advancement turn to peers for support.[4] Whatever the reason for women's powerlessness, whether it is our socialization or men's, the consequences of it are our reality.

So it is up to us to overcome our socialization in order to help men overcome theirs. For if we have been socialized to fit certain roles, they have been equally socialized to expect us in these roles. The women with the lowest mobility and the most ambition will wish to change the structure of opportunity, for they are at the base of the pyramid. The men, who are already in high mobility situations, see no reason to change the structure. In most organizations, men are still in the positions of power, with the authority to offer or withhold opportunities. As long as most men see women as not promotable, it will be very difficult for women to convince them of the contrary. What *will* convince them eventually is numbers. As more and more women attain higher-level positions, it will be easier for men to visualize women as executives and leaders.

CLAWING YOUR WAY TO THE TOP WITHOUT BREAKING YOUR NAILS

If we want to be leaders, if we are willing to fight our own socialization as well as stereotypes, prejudices, discrimination, and others' role expectations to get to the top, we need:

1. *Determination* to persevere in the face of almost unsurmountable odds.

2. *Confidence* that we have the ability to lead in the face of opposition.

3. *Energy* to work hard in spite of multicommitments elsewhere.

4. *Support* from family, friends, co-workers, sponsors, professional associations—wherever we can get it.

Is it worth it? Do we want it? Many women say no. Is it because we have been socialized to follow, or because top positions have responsibilities that do not allow time for much of anything else? To live so narrow a life does not suit many women's multiple talents for a broad range of interests. Is confinement within the family to be traded for the confinement of work? Not all women feel *compelled* to get to the top, and even many men are having second thoughts about the lack of diversity in their lives, the lack of personal and family time as they rise in the hierarchy. Each woman has to ask herself, "How high is the top for me? What prices am I willing to pay?" The answer may vary at different times in her life.

In order to know whether you *want* to be a leader and whether you have leadership potential, you must understand what leaders do, how they do it, and why. You must know the paths to effective leadership and the consequences of getting there. What are the last rungs of the *Paths to Power*, to influence, to leadership, to authority? There are two important factors in getting there: One is *wanting to* (ambition); the other is the conviction that you *can* (confidence in your competence). The first says that you believe you will be happy at the top; the second, that you believe the organization will be lucky to have you there. In other words, you are doing it both for you and for them.

When I walk into a classroom the first day of the new semester, I am aware of two feelings: the pleasure at the opportunity to teach and the conviction that the students are lucky to have me as their teacher. This feeling that the semester will be meaningful to most of us remains pervasive even though I may do less well at times and may not reach all of the students all of the time. Within confidence there can be doubt and even fears. You may start out by looking like this:

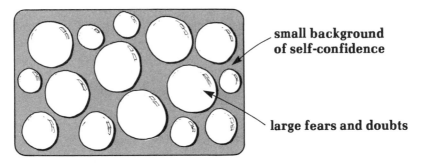

small background
of self-confidence

large fears and doubts

That's alright, because if you can chop at the fears and doubts a little at a time with even *some* success, you will begin to look like this:

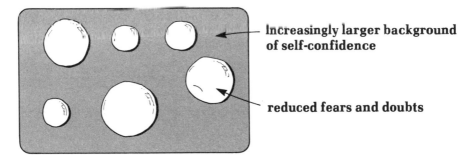

Increasingly larger background
of self-confidence

reduced fears and doubts

and end up looking like this:

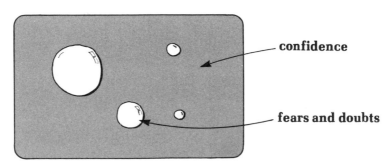

confidence

fears and doubts

The confidence grows with the opportunities to eradicate more and more of the doubts and fears. The background is permanent; the fears transitory.

"I want"

How badly do you want to be in charge? How much are you willing to be responsible for the major decisions, to be held accountable by all those below you for all the good and all the bad that may happen? What costs to your personal life do you agree to pay? Answer the following questions to know how much you really want to be in a leadership position.

1. Am I willing to let the important relationships in my life suffer because I will have to stay late, work on weekends, travel?

2. Will I be terribly upset if I am talked about or even hated by some people?

3. How will I handle being unfairly accused, misquoted, and misinterpreted?

4. Will I be able to prod the unmotivated and apply breaks to the high-risk takers when warranted?

5. Do I have my own needs for approval and for influence under enough control so that I will not be meeting these needs unconsciously at the expense of others or of organizational goals?

6. Am I willing to be results-oriented yet not compromise on the means to achieve the ends?

7. Would I fire an ineffectual but nice or needy person if it would benefit the organization?

8. Am I willing to worry about work-related problems as I go to sleep, as I wake up, and on holidays?

9. Am I ready to marry my job in addition to my husband [if married], with the consequent continuous attempt to do well by both?

The traditional male model would you say? Yes, to some extent. These are the prices men have paid for the top positions. Can many women expect to perform as effective leaders and not pay these prices? The quality of work life is much talked about today, but people are referring primarily to middle management on down. The assumption is that the top executive controls the quality of his or her life. This, however, is not so, and much effort needs to be expended to rethink the roles and responsibilities of people in top management positions.

What women might bring to these positions is a willingness to look at leadership not as the role prerogative of the leader, but as a function best invested in several qualified people. The criticism brought against women who are said to avoid success and shun power may be better understood in terms of many women's *preference* for shared leadership, shared responsibilities, shared power. Does this come only from our affiliation needs, from the belief that two heads are better than one, from a lack of self-confidence, or from our wisdom about our own limitations and our ultimate concern that people and organizations will be better served by many than by one? Empowering others is empowering ourselves. Many women have learned that the collaborative mode fulfills organizational objectives effectively.

In the meantime, whether we decide to play by the rules to obtain upward mobility or not play at all, we need to understand these rules in order to be able to make a choice. This takes us to the second necessary attribute.

"I can"

The conviction of your own competence, of your leadership potential, the "I Can" belief is equally indispensible. Ed Schein identified three types of competencies that are needed to rise to the top in a managerial career:

1. *Analytical competence:* The ability to identify, analyze, and solve problems under conditions of incomplete information and uncertainty.

2. *Interpersonal competence:* The ability to influence, supervise, lead, manipulate, and control people at all levels of the organization to work toward more effective achievement of organizational goals.

3. *Emotional competence:* The capacity to be stimulated rather than exhausted or debilitated by emotional and interpersonal crises, to bear high levels of responsibility without becoming paralyzed, to exercise power without guilt or shame.[5]

How do you know you're competent if you have not had the opportunity to test yourself? Think back upon recent events and see how you handled decision making, how you have worked with others, how you have reacted to crises. The way we behave in any one instance is often a microcosm of how we usually behave. Do you identify a pattern of responses? This pattern will give you a clue as to the predictability of your behavior. But how do you know whether you can be an effective leader? To answer this question we need to understand what leadership is.

LEADERS AND FOLLOWERS

Leadership, authority, power, influence—all these terms are interconnected. Leadership is the process or act of influencing; power is the capacity to influence; authority is the power to exercise leadership as specified by a contract between two or more parties—it is an authorization. A leader has (1) the authority to decide what should happen and who should do it; (2) the responsibility to make it happen; (3) the accountability for what does actually happen.

In any superior-subordinate relationship, the subordinate agrees to be influenced by the boss. If more authority is used than has been contracted for, the subordinate may either leave, rebel, or not do the work as expected. However, if the subordinate needs the job, he or she may comply, but with ill will. In the face of harsh authority, there are four choices: (a) to remain dependent, which may be part of an unconscious longing to be taken care of; (b) to fight, either overtly or covertly; (c) to leave; or (d) to endure, a step usually made easier with the support of others.

A leader can lead only if there are followers who agree to be led. This agreement is based on two alternatives. Either the "led" *must* follow or they *want* to follow. People who *must* follow do so because of economic necessity, fear of punishment, ignorance, or the lack of viable alternatives. People who *want* to follow do so for a potential gain, either in terms of material reward, personal fulfillment, or the achievement of a superordinate goal.

The effective leader is one whose subordinates *want* to be led as opposed to feeling coerced to do so. Followers tend to want to be led by people in leadership positions if the following criteria are present:

- The leader has the power to reward.

- The leader has a specific knowledge or skill that the followers admire or need for their own benefit.

- The leader is likable and makes the followers feel comfortable.

- The leader shares his or her power and makes the followers more powerful themselves.

- The leader represents values in which the followers believe.

PATHS TO EFFECTIVE LEADERSHIP

Leaders are effective and esteemed when they exhibit qualities that make people trust their ability to lead. McClelland and Burnham believe that effective leadership is a function of your ability to provide clear direction (as perceived by subordinates) and to foster a strong sense of team spirit.[6]

The paths to effective leadership are through *preparation, responsibility, caring,* and *creativity.*

Be prepared for both the expected and the unexpected. Being prepared for the expected means doing your homework: Know which issues will be discussed at the next meeting, have an agenda ready, be informed about inside and outside organizational matters. This will show your competence and your priorities about the issues. It will give you visibility because you'll have the needed information. But most of all, preparation gives you self-confidence.

Being prepared for the unexpected is keeping cool in a hot situation. Calm in the face of crisis, quiet in the midst of turmoil, you can cope no matter what happens because inside you are always prepared, always confident that you'll deal with whatever comes along. This feeling of preparedness in you is felt by others, who will trust the strength of your belief in yourself.

Be responsible for yourself and others. Responsibility means autonomy. It is heading projects, making decisions, being accountable, and delegating to others. A leader not only looks for opportunities for her own growth but looks for opportunities to develop her staff, to feel and be responsible for the welfare of others who are working for her. An important facet of responsibility is predictability. People trust those who have predictable reactions, on whose responses they can count and plan.

It is interesting to note that on the lower levels of management, women have trouble delegating because they want to hold on to whatever small piece of territory they have. As they move up the hierarchy, they tend to minimize the authoritative exercise of power and maximize subordinate autonomy, but they delegate mostly to males. Women need to delegate more to other women. Close attention must be paid to this; we must help each other until the time comes when women can get equal support from men.

Being responsible requires a high level of energy and time to devote to the tasks undertaken. Energy is an attribute of the powerful—it is the opposite of apathy. Charismatic people have a lot of energy and a certain intensity. Some people may feel overpowered by energy, but most will be stimulated by it. Charismatic leaders generate excitement and commitment.

Be caring about your boss, your colleagues, and your subordinates. Caring is being helpful to people. Helpfulness usually involves reciprocity. If you have helped someone, that person will help you. If people have helped you, be sure to reciprocate. Caring can be in terms of helping to accomplish a task or recognizing others' efforts. A small thing, such as saying in a meeting, "As Mary said before . . ." recognizes the contribution of a person and may be very meaningful to that person. Caring is empathy, understanding what it is like to be someone else or be in a different situation.

As a leader, you care about others, about what happens to them and how they can be helped. This does not mean that you should necessarily be *the* helper, but it is your responsibility to see that help is provided. Caring is being accessible—but do not be so available that you are unable to pursue your own work. When you take the time to explain to someone the reason for a refusal, you show concern for *that person;* when you notice and greet bosses and janitors alike, you show concern for *people in general;* when you make exceptions and do not always go by the rules, you show concern for *individuals.*

Be creative in terms of work. Creativity means risk taking: It is thinking about a problem in a new way, doing something in a different manner, being nontraditional. You are creative when you initiate rather than only respond. Problem solving is a creative activity. Forecasting is a high-risk creative act, especially if you are accountable for the results based on your predictions. Research on effective management as perceived by others found that entrepreneurship is *the* most important ingredient of a successful manager.[7] Being an entrepreneur is being innovative. One of the more important variables in creativity is relevance to the organizational objectives, that is, be creative in order to better fulfill company goals. In the process, you will reap the benefits of being the person who understands and promotes these goals.

If you are a well-prepared (confident), responsible (energetic), caring (helpful), and creative (risk-taking) woman, you are a good leader and an asset to any organization.

Transference and counter-transference

An important factor that permeates most relationships involving authority figures is called *transference*. It is outside of the rational; it is part of our unconscious. Yet because it is so pervasive and has such a marked influence on relationships, we need to understand its forces.

People in authority are the objects of transference. What this means is that people in authority are attributed qualities that may or may not be their own. Subordinates project characteristics of other authority figures onto the leader. For instance, a male employee who had a good relationship with his mother may work very well with you as leader. If not, he may attribute to you intent that may not be yours. People in power stir feelings in others, feelings that may have little to do with reality. The higher, less reachable, and more mysterious you are, the more the possibility of transference. This transference, because it is unconscious and irrational, can become polarized, in which case the person in authority is seen as either all good or all bad.

Men who do not relate well to women as individuals often project all kinds of attributes on women leaders. You may be surprised to find that some people see you as possessing the most remarkable or despicable traits. The only way to get through such transference is to become aware of it.

But leaders are not immune to transferential phenomena. You react to transference with what is called counter-transference. If, for instance, you have a very dependent employee who often asks for your approval, you may react with pleasure if you see dependence as a sign of your being needed, liked, or esteemed. You may also react with irritability if you see dependence as child-like or needlessly time-consuming. In other words, you respond to another person's transference onto you.

As a leader, it is important to understand your tolerance levels for others' behavior so that you can respond more to the person and the behavior and less to your own impulses and drives. As a woman, with your innate knowing and higher awareness levels, you have a head start at this.

LEADERSHIP AND SEX DIFFERENCES

Can women use their need for affiliation, their people concerns, and their high value orientation to be effective leaders? Studies that have correlated masculinity and femininity indices with success have found that the people who are highest in both masculine *and* feminine traits (as traditionally expressed) have the highest grade point averages, the most scientific accomplishments (as measured by being referenced by others), and the highest salaries, and are the most effective managers.[8] Current findings confirm this:

> No research evidence . . . yet proves a case for sex differences in either leadership, attitude, or style, . . . The only variable that seems to make any difference at all is the "perceived" power of the leader. Subordinates are more likely to inhibit their aggression or negativity towards a demanding person of higher status than one of lower status. Women are not seen as having organizational power, which is translated to mean either good relations with other power-holders in the system, or good upward mobility prospects. People who feel they have opportunities ahead of them seem to offer more opportunities to their subordinates to move right along with them. People who feel vulnerable and insecure are the ones who are most likely to be authoritarian and controlling leaders. The negative behavior attributed sometimes to women managers belongs to powerless people generally and not to women specifically.[9]

In a study involving thirty-six successful professional women, it was found that

> the majority of leaders seemed overtly humanistic, and directed to the 'wholeness' of the human experience. . . . This orientation was expressed strongly and intensely. An interesting consequence of this intense commitment to high standards and a commitment to working with people in meaningful ways was their equally intense dislike of "bureaucratic games." They seemed intolerant of meetings for the sake of meetings or meaningless paperwork which interfered with their commitment to professional excellence. The profile that emerges is one of remarkable strength, with the ability to think "organizationally, not just in personal terms." The commitment was to improve the work environment for themselves and others.[10]

HOW MUCH LEADERSHIP IS ENOUGH?

If too little power or powerlessness can lead to coercion, and too much power can corrupt, what then is the "correct" amount for effective leadership?

My father used to say that love is like a little bird. If you squeeze it too tightly in your hand you will suffocate it, but if you open your hand too wide it will fly away. The same applies to power. If you squeeze your hand too hard, you will suffocate your subordinates. But if you open your hand too wide, they will fly away without direction or focus, and you may lose them.

If you are a very directive leader, your people may eventually lose their sense of personal challenge, of autonomy, of control over their lives. This can lead to anger and sabotage or to apathy with low commitment to and low involvement in the task. However, there are circumstances that require directive leadership, such as in times of crisis, when members need to know that someone is in charge. Employees who feel secure can pay attention to their tasks and not bother with time-consuming participation in decision making when this is seen as inappropriate.

A democratic leader focuses on building a team or sharing each person's expertise among all of the members. From each according to knowledge, to each according to need describes this leadership stance.

When should a leader be more directive? When tasks are ambiguous, when organizational policies are unclear, or when subordinates are unable or unwilling to take on responsibilities. Leadership should be supportive when subordinates work on stressful, frustrating, or dissatisfying tasks. A leader who provides challenge by being achievement-oriented will work best with subordinates performing nonroutine ambiguous tasks. It is possible that people who choose to do routine, repetitive jobs may have a type of personality that does not respond to challenge. Participative leadership works best when subordinates prefer autonomy and self-control. We know that when people can participate in the decision making that influences their jobs, they are more committed to the results of those decisions. When people can own the decision, they have a feeling of responsibility for its outcome.

Subordinates who have authoritarian personalities prefer a directive leader who is nonparticipative. Therefore, a manager must at all times be aware of not only the types of tasks, but also of the types of employees with whom she is working.[11] How do you know when an employee is an authoritarian type? When a person is rigid, has very little flexibility, does not like change, does not respond well to the unpredictable, and relates to rules rather than to people's needs, she or he is authoritarian.

LEADERSHIP STYLE

Whether you decide to be directive, supportive, challenging, or participative, or to delegate decision making to others, you are ultimately accountable for the results. The choice of your leadership style will depend on your own preference and on the willingness and readiness of your subordinates to assume more responsibility. It is most important for you to identify your usual or preferred style of leadership. Until you identify it, you cannot make the necessary adjustments. To help you do this, complete the Leadership Style Inventory and then read the explanations of the four categories of leadership behavior.

TELL The manager identifies a problem, considers alternative solutions, chooses one of them, and then reports this decision for implementation. She may or may not give consideration to what she believes her subordinates will think or feel about her decision. In any case, she *provides no opportunity for them to participate*.

SELL As with tell, the manager takes responsibility for identifying the problem and arriving at a decision. But rather than simply announcing it, she takes the additional step of *persuading her subordinates to accept her decision*. She recognizes possible resistance and seeks to reduce resistance by her actions.

CONSULT The manager identifies the problem, *consults her subordinates for possible solutions,* and then makes the final decision. She recognizes the need to effectively cull from her subordinates their ideas to give them a sense of ownership and therefore commitment to the final decision (hers) as well as to discover other possible solutions her subordinates might know.

Leadership Style Inventory*

This inventory is designed to assess your method of leading. As you fill out the inventory, give a high rank to those words that best characterize the way you lead and a low rank to the words that are least characteristic of your leadership style.

You may find it hard to choose the words that best describe your leadership style because there are no right or wrong answers. Different characteristics described in the inventory are equally good. The aim of the inventory is to describe how you lead, not to evaluate your leadership ability.

Instructions

There are nine sets of four words listed below. *Rank order* each set of four words, assigning a 4 to the word which best characterizes your leadership style, a 3 to the word which next best characterizes your leadership style, a 2 to the next most characteristic word, and a 1 to the word which is least characteristic of you as a leader. *Be sure to assign a different rank number to each of the four words in each set.* Do not make ties. Now, total the *columns,* using only the sets numbered below in the scoring section.

1. __ Forceful	__ Negotiating	__ Testing	__ Sharing
2. __ Decisive	__ Teaching	__ Probing	__ Unifying
3. __ Expert	__ Convincing	__ Inquiring	__ Cooperative
4. __ Resolute	__ Inspirational	__ Questioning	__ Giving
5. __ Authoritative	__ Compelling	__ Participative	__ Approving
6. __ Commanding	Influential	Searching	__ Collaborating
7. __ Direct	__ Persuasive	__ Verifying	__ Impartial
8. __ Showing	__ Manuevering	__ Analytical	__ Supportive
9. __ Prescriptive	__ Strategical	__ Exploring	__ Compromising

SCORING

T	**S**	**C**	**J**
_____	_____	_____	_____
2 3 4 5 7 8	1 3 6 7 8 9	2 3 4 5 8 9	1 3 6 7 8 9

*Developed by J. F. Veiga, University of Connecticut.

JOIN   The manager defines the problem and its limitations, and then *passes to the group* (including herself as a member) *the right to make the final decision.* She feels her subordinates are capable of making decisions as good as or better than her own. She feels that human resources are best utilized by allowing them *equal* decision-making authority.

The Leadership Style Inventory describes only your *perception* of your behavior. Get feedback from others to expand on this perception. Remember also that it describes how you behave as a leader in your *current* work environment. Styles are not fixed parts of your personality; rather, they represent how you have conditioned your-

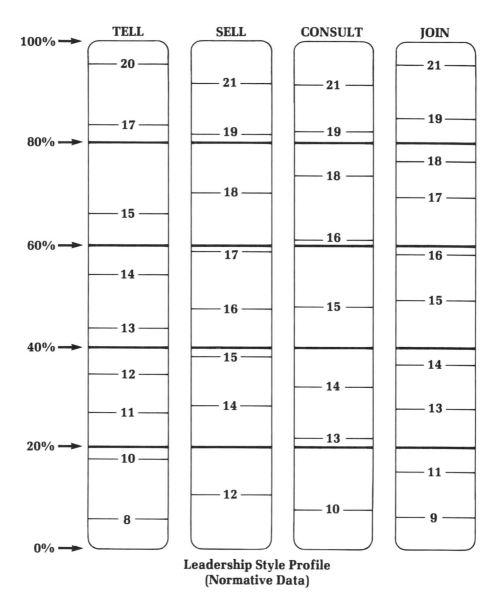

Leadership Style Profile
(Normative Data)

The above chart can be developed into a profile of your leadership style.
Shade in the area which corresponds to your score on each dimension.
For example, if you scored 15 on the TELL scale, then shade the area up to
the 15 under TELL on the above chart. The ruled-in percentile provides
you a way of comparing yourself to others who have taken the inventory.
The percentiles are keyed to style scores to indicate the number of people
who scored below a particular score. For example, a score of 15 on the
TELL style means you scored higher than almost 65 percent of the people
tested.

Possible Ways for Leaders to Behave

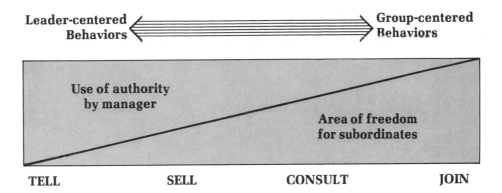

Leader-centered Behaviors ⟷ Group-centered Behaviors

Use of authority by manager

Area of freedom for subordinates

| TELL | SELL | CONSULT | JOIN |

*The Tannenbaum and Schmidt Leadership Model. J.F. Veiga, University of Connecticut.

self to respond to your current work situation (how you have chosen to cope to meet your needs). Therefore, focus should be on understanding and diagnosing why you chose the style(s) you did and on evaluating whether you made the best choice. Each one may be correct in specific situations.

LEADERSHIP ISSUES

"The world is divided into those who look and those who are looked at."* If you are a leader in a power position, you are looked at. If you are a follower, or in a powerless position, you do the looking. What does this mean? Leaders are visible: Their actions are constantly commented upon; their decisions are held up to public scrutiny; they have no hiding place. They are admired, hated, misunderstood, misinterpreted, sought after, avoided, tested, deferred to, and agreed with.

One of the gravest dangers threatening people in top positions is the lack of information. Subordinates, wanting to ingratiate themselves, will only report good news; hard truths will be kept from the top executive, who then acts on misinformation and lack of data. It is critical for an effective executive to keep seeking new data and to be open to unpleasant information, be it about her leadership style or the consequences of her decisions or policies.

Leaders are also subject to counterdependence from people who react negatively to anyone in authority. A woman may be more prone to elicit such counterdependence because of her gender. In this case, anything she says or does will be seen negatively—it has nothing to do with style or competence. It is important to differentiate between taking the blame and leaving it out there where it belongs, in the prejudicial attitudes of some subordinates.

There is every likelihood that a psychology of entitlement will prevail in the 1980s. This means that more employees will feel entitled to more interesting jobs, to a more satisfying life-style, and to more personally challenging and autonomous

*Dean Charles Warden, UNH class lecturer, 1978.

work. Some of the key problems will center around worker alienation, low commitment to job, and resistance to authority. People will no longer accept transfers as easily as before; dual-career couples will have to be taken into account; child care will become a bigger issue. Values are slowly changing in the world of business and they are shifting in the direction of women's basic orientation. So while our male colleagues need to make an effort to encompass these new values, we feel more and more at home.

What are some of these changing values?

- Placing a greater emphasis on collaboration as opposed to competition.

- Paying attention to process and not just to task.

- Trusting people.

- Sharing power.

- Being authentic as opposed to playing games.

- Appropriately expressing feelings rather than shutting them off.

- Viewing people as whole persons and not just in terms of a job description.

- Accepting and utilizing individual differences instead of resisting or fearing them.

- Showing many personality facets rather than only those related to work.

Women have already arrived where most men are just beginning to get to!

NOTES

1. William E. Reif, John W. Newstrom, and Robert A. Moncka, "Exploding some myths about women managers," *California Management Review* 17- 4 (Summer 1975).

2. Agnes K. Missirian, "The Female Manager as a Shelf-Sitter," *Human Resources Management,* Winter 1978, p. 29.

3. Jean Baker Miller, *Toward a New Psychology of Women* (Boston: Beacon Press, 1976).

4. Rosabeth Moss Kanter, *Men and Women of the Corporation* (New York: Basic Books, 1977).

5. Edgar H. Schein, *Career Dynamics* (Reading, MA: Addison-Wesley, 1978), p. 31.

6. D. C. McClelland and D. A. Burnham, "Power is the Great Motivator," *Harvard Business Review* 54–2, 1976.

7. David Ley, "An Empirical Examination of Selected Work Activity Correlates of Managerial Effectiveness in the Hotel Industry Using a Structured Observation Approach," doctoral dissertation (East Lansing, MI: Michigan State University Graduate School of Business, 1978).

8. Janet T. Spence and Robert L. Helmreich, *Masculinity and Femininity* (Austin, TX: University of Texas Press, 1978).

10. Jyotsna Sanzgiri, "Professional Women: An Experiential and Theoretical Analysis," doctoral dissertation (Pittsburgh: University of Pittsburgh, 1977).

11. Adapted from Charles N. Greene, "A Causal Interpretation of Relationship Among Pay, Performance, and Satisfaction," paper presented at the annual meeting of the Midwest Psychological Association, Cleveland, May 1972.

SELECTED READINGS

Drucker, Peter F. *Managing for Results*. London: Pan Books Ltd., 1964.

Fiedler, Fred. *A Theory of Leadership Effectiveness*. New York: McGraw-Hill, 1967.

Gibb, C.A. (Ed.). *Leadership: Selected Readings*. New York: Penguin Books, 1969.

Hersey, Paul, and Blanchard, Ken. *Situational Leadership*. San Diego: Learning Resources Corporation, 1976.

Horner, Matina. "Fearing Success." *Psychology Today*, November 1969.

Irish, Richard K. "Ten Questions to Ask Yourself if You Think You Want to Be Boss." *Glamour*, October 1976.

Stodgill, Ralph M. *Handbook of Leadership*. New York: Macmillan/Free Press, 1974.

"Women in Business: A New Look." *Journal of Contemporary Business*, Winter 1976.

Chapter 12

WHAT AM I?

Somewhere between always giving to others
and always keeping it all to myself
I stand.

Somewhere between only caring for others
and only caring for me
I live.

But when I am only for others
I ask,
Who will be for me?
And when I am only for me
then what am I?*

*Adapted from Hillel, Aboth 1:14, *The Talmud of Babylon.*

210

Top Executive

Influence power—the top! This is the place so many people dream of. The pinnacle of the pyramid, the last rung of the ladder, the top of the hierarchy, the highest level of management! Arrival here is proof beyond doubt that you are successful, valued—a leader. Your pay range may be in the six digits; you may have servants, a chauffeur-driven limosine, and perhaps a private jet at your disposal. Yes, it is glamorous and exciting—and there are practically no women there! As of this writing, there are no women CEOs of the *Fortune* 500 companies.

Because of affirmative action, more women are now hired for entry-level positions, but they remain at the lower rungs of management. Tokenism is the sharp issue at the top. One position might be opened at the top for a woman ("the woman's slot"), but once filled, all other available positions will be offered to male candidates only. "We already have a woman here" is an oft-heard statement, the implication being that no more women need to be even considered.

As women swell the middle levels of management, that *one* slot will foster strong competition among the women who vie for it. Therefore, it is important for companies who believe they are fulfilling their EEO requirements by reserving *one* top position for women to become aware of the discrimination this practice perpetuates.

I have often seen the following posted on office walls:

Anything a woman does
Must be done twice as well as a man
To be considered half as good.
Luckily, this is not difficult!

It may be true that it is not difficult for many a woman to be twice as good as a man, but it is also true that even if she is, she will indeed be seen as only half as good.

It is difficult at best, under these conditions, to remain ambitious and goal-oriented: The goals are so much harder to reach for women than for men. And because they need to work harder, women give up more than men in terms of their personal time and a private life. In cost-benefit terms, women report that the costs in time, outside interests, and family, exceed the benefits of the top-level jobs. Men reported that the benefits outweigh the costs.[1] In other words, women must compete harder and give up more in order to have a lesser chance for an executive position.

A woman who aspires to be top executive needs to be more talented, more capable, and more persevering than most men because she encounters more roadblocks on her way up. But just like men, she needs to have the specific competencies in her field plus the political skills to get head. In talking with over thirty women in the upper echelons of their organizations, I asked them five questions: (1) What blocked their careers? (2) What helped their careers? (3) What are the benefits of being at the top? (4) What prices are they paying? and (5) What advice can they give to aspiring women? I have divided the most common responses to questions (1) and (2) into two categories: those barriers and aids found internally and those found externally.

Is it all worth it? Only *you* know. Does *more* responsibility at work give you satisfaction or does "growing" outside your career fulfill you better? Often, the climb to the top sacrifices relationships. You may have that choice to make. There are no right answers; there are only *your* solutions, right for only you. Before you take the next step up, ask yourself if where you are going is really where you want to arrive!

Internal barriers	External barriers
Lack of insight into the workings of the white male culture*	Male colleagues' lack of perception of my competence and capabilities
More than half the pressures were self-imposed	Men's uneasiness and discomfort working with women
Belief in the myths: (1) If you fail once you won't be given a second chance (2) You can't say "No"	Male belief that even if women could problem-solve, their clients would not trust them
Lack of exposure to business	Had to *over*-demonstrate my competence to earn the team members' respect
Not knowledgeable of the informal power-structure—took four years to learn it	No mentors available
Too task-oriented	Quota mentality: only one top slot available for *all* the women
Lacked initiative	The first woman in the firm
Did not know for *sure* what I wanted to do	
Not assertive early enough	
Always *grateful* to have a job	

*White male culture refers here to vertical bonding (men socializing up the hierarchy), paying your dues (doing what needs to be done even if unpleasant), and leaving when the time is ripe.

Help found internally	Help found externally
Sheer gut and survival power	Recognized as being good
Competence to sell what others value	One male colleague explained often what was going on
Persistance and determination	Women peers were a support group
Learned to market my skills, in particular a financial background	Had an advocate
Love my career; cannot conceive of not having one	
Ability to demonstrate results (results-oriented)	
Without role models, developed my own style of neither denying my sexuality nor flaunting it	
Developed self-confidence through testing it	
Developed attitude that "business" is *not* a dirty word	

WHAT DOES A TOP EXECUTIVE DO?

A top executive usually works through an executive committee, which consists of herself and her vice presidents. She reads briefs and summarized reports but must rely on information obtained from others for her assessment of various situations and base major decisions on what is reported to her. Therefore, mutual trust becomes a critical factor. Since policy decisions are projections of the future, an executive deals with what is not yet known. Because there is no available information that can reliably predict outcome, she relies on her experiences; on the study of political, economic, and sociological trends; and finally, on her own intuition.

Of course, the ultimate accountability rests with the top executive: the buck truly stops at the top. Middle managers can always say, "If I were in charge, I would do things differently." The executive cannot say this. She often feels powerless when things have not turned out as planned. Not carrying out executive policies or decisions and carrying them out other than as planned are two forms of veto power that all the people below the top have. The only veto power the executive has is to fire. To guarantee correct implementation of policies, the CEO must involve as many people as possible in the decisions that affect them. But the executive rarely communicates directly with *all* the employees; instead, information is passed through the various management levels both up and down. The executive must constantly check that the information she gives and receives is accurate and complete.

Benefits	**Costs**
Power	No children
Influence	Not being 'taken care of'
Money	Loneliness
Autonomy	Less well-balanced life
Official recognition	No time to read or travel, and not willing to devote any time to it.
Affluent life-style is ego-satisfying	No time for tennis and squash
Ability to make things happen that are important to me	Not putting enough time into the marriage
Attitude that you are as good as you think you are	Narrow pursuits a problem, because in order to make good management decisions and meet my responsibilities to the company, the shareholders, and the community, it is important to be well educated
Having protégées and watching those you counsel get better	
Intellectual stimulation	
Dealing with issues in new areas can be heady	"The buck stops here"
Working with a stimulating team made up of challenging people	Working harder with more pressure
	Less free time and less flexibility
	More time spent with others on a professional basis
	No privacy, always responding to others
	Need to work harder and longer than men do, which will probably never change

A day in the life

Let us look at a day in the working life of a top executive. You will note that it does not seem very substantive, for at that level, the job is almost all *process*.

Morning—

▰ Scans newspapers. If there is an item pertinent to the organization, asks for an assessment of the situation by the appropriate person.

▰ Reads various reports, for instance, the steps taken by the firm to cut energy consumption.

Advice
(From me as I am *now* to me as I was *then*)

Not everyone in authority is invincible

You're not the only one suffering

Really think through what you want to do and plan accordingly

Relax more and be less anxious

Develop good skills or great competencies—be *super* at something

Learn to assume about 60 percent of the blame; the rest is outside yourself

Stop looking at men as fathers and gods; see their warts, see their limitations as fellow human beings

Have a stronger self—this will help you take more risks

Do not take things personally—be *objective*

Become an astute observer of how decisions are made and who makes them

Try to get a position with the star player (the one to whom top management seems to be paying the most attention)

Do not be reticent about your skills and talent and do not wait for praise, but do not blow your own trumpet too hard, either

Hook up with a professional women's network for advice, support, and the chance to ask "dumb" questions; you will be lonely for the next ten years

---

- Calls lawyer to discuss possible take-over threat by large conglomerate.

- Talks with vice president for labor relations about strategy in dealing with the strike rumors in one of the plants.

- Comes a few minutes late to executive meeting that has been called to look at the feasibility of shifting from functional management to product management.

- During meeting, gets interrupted by urgent phone tip that the government is about to order a massive recall of a competitor's defective product. Calls an emergency meeting to assess how to take advantage of this situation.

- Dictates letters and memos.

Afternoon—

- Has lunch with the president of a major company that may become a customer. On the way back to the office, stops at cleaner to pick up dress for that evening's dinner and speech.

■ Negotiates with the bank committee the financing of a possible merger with a company in an allied field.

■ Meeting with a local consumer group.

■ Talks with vice president for production about an unexpected drop in the demand of a product during the last quarter.

■ Returns to office to find a former classmate waiting, who wants a contribution to a college fund-raising drive.

■ Receives request from a broker to find the man's daughter a summer job with the company. Passes this to administrative assistant.

■ Sees reporter, who is checking on story about price fixing.

■ Calls husband to remind him that she won't be home for dinner.

■ Attends local community-service organization's board meeting, which she chairs. Reads latest report of antitrust suit on the way.

■ Leaves early to deliver opening speech at regional sales conference.

■ In briefcase sits report of audit committee, still unread. Next day the quarterly report will be out, which means seventeen people will call to ask why stocks are down 10 cents a share, which means going to work early.

Although this may or may not be typical of the work load of the top executive of a large corporation, it *does* indicate typical organizational issues.

The following chart shows the differences between the viewpoints and responsibilities of managers and executives.

At the top levels of management, what needs to be done is dictated by the setting specifically and by the larger organizational issues generally. Whether the company is in manufacturing, service, government, or health and education will influence the types of issues faced by the executive.

A *manufacturing operation* is hierarchical in nature. Differentiation is by level of skills. Technical backgrounds are valued and the product is most important. Balancing stability with change will be the critical dynamic. The issue for an executive might be to manage the tension between maintaining low-cost standardization to ensure a high volume of productivity and remaining competitive by training personnel to use new technologies.

Service organizations, such as retail, banking, and insurance companies, have a large low-skilled labor pool and are usually divided on a regional basis, with an emphasis on marketing. Here the issues will be around recruiting, training, and motivating workers with low morale. High turnovers and cost controls must be dealt with. An example would be the problems of using young people in supermarket jobs.

Government agencies employ civil servants selected and promoted by a point system. Most agencies are rules-ridden. Some of the issues faced by an executive would include acquiring enough flexibility to increase effectiveness yet remaining within what is allowable. It is difficult to maintain creativity in an environment constrained by rules.

Health services and educational and cultural institutions usually have a hierarchy with three levels: the professional people, the support staff, and the administration. One of the issues would be to retain the separate identities of these groups yet promote communication between such people as doctors and nurses or teachers and school administrators.

Besides setting, another factor to be taken into consideration is the size of the organization. The larger the company, the less visible the CEO and the more she must rely on the reports of others, with little first-hand information. Trustful relationships with her staff are critical.

Top executive	Manager
Oriented to the present and distant future	Oriented to the present and immediate future
Makes policy	Implements policy
Integrates the various functions	Responsible for functional management
Represents company to the community	Represents company to its employees
Relates to relevant external environment	Relates to internal constituency
Secures commitment to purpose	Secures commitment to job
	Flexibility to modify as appropriate, mistakes are less costly.
Policy decisions are so complex that incentive to modify is minimal over the short run	
Concerned with expanding, raising money, investing	Concerned with getting needed people and resources and with how to use them
	Personally intervenes
Manages the policy-making process; is less directive	Supervises and coordinates current operations (manages specific operations)
Plans for future operations (general management)	
	Delegates tasks to lower-level managers
Allocates decision making and authority to functional areas	

(continued)

A top executive also deals with some of the following issues, with which managers are not specifically concerned. The differences may be at times a matter of emphasis given to certain responsibilities.

Negotiates with labor unions

Must know different aspects of the laws and government regulations; takes responsibility for ensuring compliance

Pays attention to public service

Concerned with the moral dimensions

Must have courage to avoid shoddy solutions

Clarifies goals and defends them against outside attacks or internal erosion

Decides strategic alternatives in view of marketplace and company resources

Continually monitors suitability of company purpose

Needs personal magnetism

OPPORTUNITIES FOR INFLUENCE

The power at the top lies in the opportunity to make things happen according to your personal value system. Besides being able to decide on the direction of the organization and on how much authority to invest in others, the top executive has social and moral responsibilities. Acting upon some of these opportunities can make a large difference in the lives of many people and in their environment. Contributing to charities, supporting public television, funding research projects and educational institutions, supporting programs to enhance the quality of work life, generating equal opportunity, and paying attention to pollution and other environmental hazards are typical concerns of an executive. The decisions around these opportunities must be weighed against the cost to the company. Once made, commitment to these programs must be elicited from the interested parties.

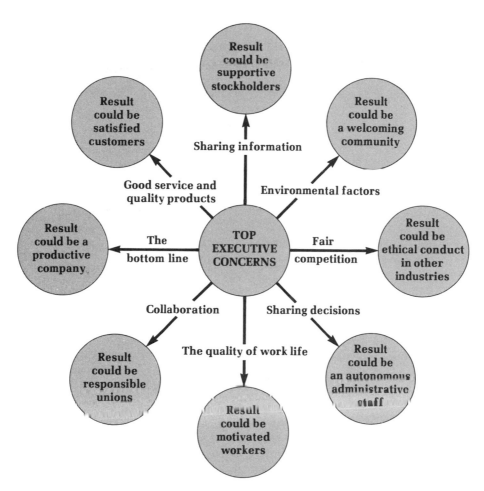

A central issue for the people in power relates to social consciousness. Responsibility to this issue is determined not only by the individual firm, but also by the government. Presently the government sets many standards for organizations. The CEO must decide how to balance the need to make as much profit as possible with the need to be cognizant of the possible effects of her company on the physical and social environment. Hazardous waste, air pollution, and the noise level are becoming more and more important concerns.

Whereas a middle manager focuses her energy to understand what is going on *inside* the organization, the top manager focuses her energy *outside,* scanning the environment in order to understand how it and her organization impact each other. She needs to know the technological changes that are occurring, the national political climate, the attitudes toward monopolies and tariffs, the legal constraints, the diverse climates for foreign investment, the areas that need more lobbying. She will have to forecast business cycles, understand economic recession forces, and know the law. She needs to know about interest rates and currency exchanges, to be aware of the availability of monies and resources.

Values change in our society, demographics influence a firm, and the norms of the society alternately provide threats and opportunities. The CEO needs to know the weaknesses and strengths of her organization in order to protect it and take advantage of the constantly changing economic, sociological, and political climates. The head of a large firm may be an important force in politics, in applying the law as intended, in making a difference for thousands of people.

EXECUTIVE CHARACTERISTICS

By the time a woman has made it to the rarified atmosphere of the top, the primary issue will not be her acceptance in the organization. Her problems now center around her board of directors, trustees, government leaders, clients, top executives of other companies, bank officials, politicians, stockholders, and the heads of charitable institutions. However, the very fact that she has made it means that she has survived all the pitfalls of the upward climb and has therefore learned to strategize along the way.

At this level, there are no job descriptions, only issues to be dealt with. A candidate for an executive position must have a background of proven successes. She must be seen as capable of meeting the expectations of various constituencies, such as shareholders, directors, company employees, and the buying public. She must be results-oriented, watching that bottom line, and able to integrate functional areas within the organization and to remain in touch with the community and with relevant dimensions of the world at large—all this in the face of the uncertainty idiosyncratic of the top.

The knowing base

There are no directives on how best to prepare yourself. This level of leadership is not learned in a systematic way, but opportunities for learning it exist throughout your move up within the organization. It is a combination of experience, good judgment, and ambition; it is knowing as well as knowledge. More than ever, at this level, you need to trust your intuition, to risk actions based on predictions of the future, to make decisions based on others' perceptions and recommendations.

It is also important to note that although until recently most CEOs came from marketing, today most seem to be coming from finance. The prediction is that finance will still provide most executives for the immediate future. This should indicate that one of the best routes to take is through accounting, budgeting, and financial management.

There is also the element of luck. What is luck? It is not only chance, it is also *creating* the opportunity, *recognizing* it when it is there, and *taking* it when it comes. Luck is being at the right place at the right time, but location and timing are to some extent under our control. If you are ambitious, your eyes and ears should always be open for the slightest indication of possible opportunities. I think the people who always seem to be lucky have merely developed a good nose for sniffing out the next opportunity—they use their knowing.

The knowledge base

The knowledge base deals with understanding the total system, known as *systems theory,* understanding organizational development, and understanding the dynamics of conflict, especially intragroup conflict.

Systems theory   An organization is a system comprised of many parts existing within a wider context. Systems theory deals with the boundaries of the parts and the exchanges between them, both between the various units of *one* organization and between an entire organization and its environment. Organizations have three main components:

(1) *Objective* People belong to organizations in order to work together to fulfill a purpose they hold in common, because one person alone could not fulfill this objective.

(2) *Structure* An organization's structure divides the various tasks and responsibilities among various people in order to permit them to cooperate with each other to achieve the organizational purpose more easily and/or quickly.

(3) *Process* The organization must have a process by which disagreements are handled, conflicts are managed, adjustments are made, communication patterns established, and leadership formalized.

In other words, an organization needs a process in order to establish a structure that will fulfill the purpose. The process is *how* you do it; the structure is *what* you do; the objective is *why* you do it.

The Open System   An organization does not exist in a vacuum. It exists in another, larger system, which has social, cultural, material, economic, and political components. Since the organization exchanges freely with the outside environment, it is called an open system.

The basic open system model is that of an input-output system. The organization functions as the converter from what is put in, such as raw material, and what is put out, such as a finished product. (In an open system such as a hospital, the raw material put in is sick people, and the finished product put out is well people. Or again, what is put in are doctors in training, and what is put out are experts in their field.) However, more has to be put in than comes out, because some of it has to be utilized by the system for its own maintenance. For instance, it needs money to pay for personnel, to buy equipment, and to finance research that may or may not have an immediate visible output. Herein lies the danger for all systems: that they will not have enough input for maintenance and therefore will go bankrupt, disintegrate, die.

This is the tendency for all systems; it is called entropy. Very few systems have lasted more than a century or so. For a system to withstand entropy, it must continually adapt to both its internal tensions and needs and its external exchanges and demands, thereby attaining equilibrium. This state is called homeostasis; a dynamic homeostasis indicates a changing equilibrium that adapts to the environmental pressures, without losing the identity of the system.

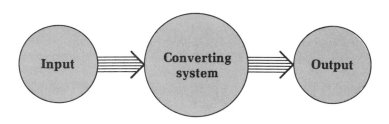

Systems can learn to control their development and remain in equilibrium through a *communication loop,* a system of information feedback that allows the organization to keep adjusting to its environment. New behavior will elicit more feedback, and this continuous sharing of information will allow for maximum growth.

Organizational development What is organizational development (OD)? Richard Beckhard gives the following definition: "Organization development is an effort (1) planned, (2) organization-wide, and (3) managed from the top to (4) increase organization effectiveness and health through (5) planned intervention in the organization's 'processes,' using behavioral science knowledge."[2] Organizations, just like people and groups, go through developmental phases. They have a tentative beginning, when they feel themselves out; a period of growth, characterized by excitement and mastery; then a leveling period, when things settle down.

In terms of control, OD deals with diagnosing the system in order to improve it. Organization behavior (OB) is diagnosing people and groups within the system in order to improve their effectiveness. To cause change when you are top management, you need to understand both OD and OB and how they apply to systems.

For peak performance, all the parts of any organizational system need to interact smoothly and well. When something goes wrong between groups, or when a problem within one unit becomes chronic, the top executive must use her knowledge of OB, OD, and systems theory to correct the situation. Although there are times when consulting a specialist is appropriate, an executive should be able to deal effectively with most problems that might arise.

Sometimes diagnosing or locating the problem is difficult; sometimes finding the correct solution or implementing the decision is harder. The locus of the problem could be in a group that is consistently ineffective, or between several groups vying for power, or in ineffectual leadership, with a manager feeling overloaded, not respected, or indecisive. The problem could also be in the *total system:* The labor market or the physical plant could be inefficient or a general climate of malaise or anxiety could exist. When an executive is unable to cope with these problems, it is time to call in a consultant.

Intergroup conflict   Just as in the two-person work relationship, groups can work together in three different ways. One is reciprocal interdependence, in which no group can function on its own. The second is serial interdependence, in which one group cannot start its work until another has finished. The third is pooled interdependence, in which the various units all contribute to the company's productivity but do not work directly with each other. The issue between groups is always one of collaboration versus competition. Even though collaboration may seem like the preferred mode in all situations, this is not necessarily so. A certain amount of competition could motivate a group to achieve its objective either better or faster than another group. However, competition generally increases productivity only when the groups concerned work together in a pooled interdependence relationship.

Controlling competition   Competition, when carefully controlled, can foster strong group identity, which feels *good* and can spur a work unit to better performance as members of a team. It can also engender a winner/loser attitude that can be debilitating.

If there is a winner, there must be a loser—a unit may feel resentment toward the victorious group, dejection over its own loss, and possibly even anger at its own team members, as each person tries to displace the blame for failure. Eventually, the group may feel so demoralized that it just gives up. Losing does not necessarily result in apathy—it could instead act as a catalyst to make the group outdo itself next time around. However, this can be an uphill climb if the winning group's attitude is not helpful.

The members of the winning group will eventually perceive themselves as being *better* than the other groups. This can lead to overestimating their own competence, overvaluing their leaders, and distorting any information about other groups to fit the perception of their being less important. This limits communications, producing more mistrust and misconceptions. This can become a vicious cycle if competition is escalated. This is not unusual and a top executive must constantly be aware of the problems competition can foster.

As top executive, what should concern you most when a conflict arises? Should you worry more about the task and whether it is accomplished well enough to fulfill the company objectives? Or should you put the people uppermost and strive for their contentment? Depending of course, on the particular conflict, paying attention to both is the best course—try to maximize productivity without sacrificing personnel satisfaction.

An executive is often faced with the dilemma of how to resolve group conflict: Should she impose a solution, let the parties figure it out by themselves, or actively work with them to find a mutually acceptable solution? How the executive chooses to manage conflict will depend upon her leadership style, the people involved, the situation, and the available time.

An executive who is responsible for the working of various competing groups can use several tactics to induce better collaboration. One is to help the groups recognize and then accept common goals. Another is to provide as much communication and interaction between the groups as possible, for that will help the demysification process. The more frequently and the more openly information can be shared between groups, the better they will be able to collaborate. If the executive gives them a common task, the groups become that much more cohesive.

One of the ways to help solve extremely dysfunctional group competition is to ask each group to jot down anonymously their perceptions of the other group. Then meet with the concerned groups to share their perceptions. Some are based on data; most, however, are based on fantasy. It is critical for the executive to be as impartial as possible. Sometimes, however, taking sides is unavoidable, so you must be very clear of your reasons for having made a choice. If you have indeed been neutral, confirm that position publicly with both groups.

The Give-or-Take Model I would like to suggest using the following model to handle conflict.

The Give-or-Take Conflict Handling Model

Give Nothing	Give and Take	Give Half	Give All	Give Up
("I'll never give in")	("Let us problem-solve together")	("Let us compro-mise")	("I'll give in to you")	("I'm leaving it unresolved")
		Possible Reactions:		
Angry or relieved	Satisfied or bored	Accepting or unfinished	Victorious or guilty	Frustrated or pleased

All five ways may be appropriate in different situations. For instance, you may decide one group should *give nothing* when you have a crisis on your hands and someone must make a decision immediately. It also may be your stance when you are sure one group is right and you want to push that idea at all cost. To *give and take* is to have the groups problem solve together. It is the most time-consuming mode, but it may be important when each group's participation in the decision-making process will ensure commitment to the results, or when you need the expertise of people from each group to make the best decision.

The third way of dealing with conflict is to *give half.* It can be a negotiating stance, and each side gives up something on an issue in order to achieve a workable resolution. When there is no hope of achieving a totally mutually satisfactory result, this mode is appropriate. The fourth conflict mode is to *give all.* This is useful

(a) when the particular issue is not terribly important to either group; (b) when it is a no-win situation that doesn't warrant a lot of energy; or (c) when one group decides that, by giving in this time, the other group will owe it a favor. The last mode is to *give up*, to leave the conflict unresolved. Withdrawal may be appropriate when the issue is not worth fighting over, or is too hot to handle, or is one that a group would prefer to avoid. You may direct the concerned groups to other means of solving their conflict that do not involve you, or you may decide to simply shelve the issue until you gather more information or tempers cool down.

No matter which mode you follow, you must be aware of the consequent emotions your action may arouse in those involved. For instance, if you decide that one group should give all, the other group may feel angry at you for not considering its needs, and the victorious group may feel guilty at winning even though it expresses satisfaction. A give-and-take mode, on the other hand, may make some people impatient at what they consider a time-waster, while others will be quite content. In other words, expect unexpected reactions.

Another way of approaching it would be to define the problem by identifying the differences of opinion concerning it. Encouraging the legitimacy of those differences and the feelings attached is an important first step. Then you can determine if there are any agreements or similarities of opinion and feelings about the problem. Once there is agreement as to what the conflict is and how the parties feel about it, then it is possible to see whether either party is willing to give up, give all, give half, give and take, or give nothing.

Each conflict mode has virtues and pitfalls. The important thing is to realize that no single mode is always appropriate, but that effective conflict management is situational. You may assume that a give-and-take mode is naturally superior to a give-nothing. It is so *only* when time is not critical and commitment to a decision is. A person's usual way of handling conflict reflects a preference which may be a pattern. This chart is to help you broaden your range and enlarge your repertoire of responses to conflict situations.

Consensus   To reach a decision, people do not necessarily *have* to agree! But people *must* have the opportunity to vent their thoughts and feelings about the issue. Consensus can only be achieved when all concerned feel that their opinions have been heard. Then people can agree to disagree—they can go along with an action in which they do not totally believe in order to help the group's progress. Because women tend to have good interpersonal and verbal skills and are good process observers, we should be able to confront conflict in a group in terms of the problem as we see it, noting people's reactions to the problem and getting agreement on its resolution or management.

Managing conflict with compassion and justice while keeping in mind company goals is not easy. An executive must be aware of the physical environment, organizational objectives, and the human factor, and her actions must have integrity, representing her value system.

PRIVILEGE AND OBLIGATION

A woman with influence power has a large responsibility toward the women still in dependent power positions. A woman executive has followed the *Paths to Power* the way no men ever have or will and she can now share her experience with other women in order to facilitate their upward mobility.

She must become a sponsor to younger women and to late-entry women, recognizing the special nontraditional competencies many of them possess, identifying the roadblocks and providing the guideposts. A woman at the top has both the privilege and the obligation to help other women make it. By providing opportunities she builds a strong team, which will not only benefit many others, but, in the long run, also herself.

Hopefully, there will come a time when multicommitments and a humanistic value system will not be incompatible with success. We need to retain our femininity in the highest rungs of management, in our styles of leadership. Women can remain sensitive, aware, warm, and caring and still be all the things an executive needs to be and do all the things that need to be done.

NOTES

1. Ann Harland and Carol Weiss, "Moving Up, Women in Managerial Careers," ongoing research (Wellesley, MA: Wellesley College Center for Research on Women).
2. Richard Beckhard, *Organization Development: Strategies and Models* (Reading, MA: Addison-Wesley, 1969).

SELECTED READINGS

Alpert, Judith, and Richardson, Mary Sue. "Conflict, Outcome and Perception of Women's Roles." *Educational Gerontology* 3(1978):79–87.

Beckhard, Richard, and Harris, Reuben T. *Organizational Transitions: Managing Complex Change.* Reading, MA: Addison-Wesley, 1977.

Cohen, Allan; Fink, Steven; Gadon, Herman; Willits, Robin; with Josefowitz, Natasha. *Effective Behavior in Organizations.* Homewood, IL: Richard D. Irwin, 1976.

Frost, Peter J.: Mitchell, Vance F.; Nord, Walter R. (Eds.). *Organizational Reality, Reports From the Firing Line.* Santa Monica, CA: Goodyear Publishing, 1978.

Gibson, James L.; Ivancevitch, John M.; Donnelly, James H., Jr. *Organizations* (rev. ed.). Dallas, TX: Business Publications, 1976.

Klein, Donald C. "A View from the Park Bench on Organizational Diagnosis." *Social Change* 7-3(1977).

Kundsin, Ruth B. *Women and Success: The Anatomy of Achievement.* New York: William Morrow, 1974.

Mednick, L.; Tangri, S.; Hoffman, L. (Eds.). *Women and Achievement.* New York: Halsted Press, 1975.

Robbins, Stephen P. *Managing Organizational Conflict: A Nontraditional Approach.* Englewood Cliffs, NJ: Prentice-Hall, 1974.

Schein, Edgar H. *Organizational Psychology* (second ed.). Englewood Cliffs, NJ: Prentice-Hall, 1970.

Sisk, Henry L. *Management and Organization* (third ed.). Cincinatti, OH: South-Western Publishing Co., 1977.

Tannenbaum, Robert, and Davis, Sheldon A. "Values, Men and Organizations." *Industrial Management Review,* Winter 1969.

Wren, Daniel A. *The Evolution of Management Thought* (second ed.). New York: John Wiley & Sons, 1979.

Epilogue

PLUMS AND PRUNES

Your skins are taut,
Your faces smooth
    like fresh plums.

My skin is wrinkled,
My brow furrowed
    like a prune.

Prunes are sweeter!

Re-Tire to Re-Start

Look back a moment. You have been a worker, an entry-level woman, a supervisor, a manager, and perhaps even an executive. You have been in dependent power positions and you have known influence power. You have used well your knowledge and your knowing and now you are nearing that period in life most of us must face: the senior years.

Our society tends to disparage the elderly and overvalue youth. There are many negative myths about old age. We somehow tend to think that, automatically at the age of 65, people lose their vitality, their memory, their sex drive, their ambition, and their physical and emotional flexibility, among other things. Elderly women are doubly discriminated against: They suffer from female *and* age stereotypes. Consequently, as soon as women turn "that certain age," it is assumed they will bow out gracefully of their own accord. If not, there is always the law to effect what convention could not.

The new mandatory retirement age of 70 is a positive step toward the once-common recognition of all that age has to offer—wisdom and skill earned through great experience. Still, retirement is inevitable, and it can be painful. For many people it means partial or total individual powerlessness. For a woman who has worked most of her life, this transition can be very difficult. Retirement can also mean cultural estrangement and/or social isolation: Relationships formed at work are often primary social contacts, and retirement can break these bonds. Another major change is in income, which is almost always lowered. These factors can combine with the loss of status or identity that was offered by occupation to leave people feeling purposeless, therefore worthless, and some tend to just give up.

RETIREMENT

It doesn't have to be that way, however. If you plan to retire, *plan your retirement*. Think about it ahead of time. Use your financial knowledge to prepare for the income drop—especially if you will have to change your residence—and to budget your smaller income. Use your forecasting skills to anticipate and forestall any particular problems you might expect to encounter. If you worry about unstructured time, schedule it to fill up your day. Keep in mind that anything you do will have more meaning for you if it can have meaning for someone else, too, or if the activity or its results can be appreciated by others. Use your talents, keep your mind active. *Stay healthy* by paying attention to your mind as well as your body. Remain interested in outside activities, watch your diet, and exercise. The more active you are now, the more active you will remain!

If you plan ahead, the transition to retirement can be very smooth and you may escape most of the stresses inherent to this change in status. Make as few other changes as possible during this process, since *any* change— even a positive one—causes stress. For example, try not to move to a new apartment the same month you retire.

Retirement can be wonderful if you see it as *starting* something rather than *stopping* something. The life expectancy for women is 84; if you retire at 65, you have a full twenty years in which to expand your horizons—to be, do, read, see, go to, and write about all the things for which you never before had time. You can even start a new and completely different career. Just because you are 65 does not mean you cannot take courses, learn something new, get a degree. And you will still have enough time to *use* your new knowledge, to *practice* your new skills.

Life might be compared to a long sea voyage, with each of us in an individual boat. Some people were born to be in swift motor boats, others were born to be in sailboats that veer with the changing winds. Still others have to work hard paddling their canoes upstream, and some have no oars at all—they just remain stuck, going nowhere. Women's boats, of whatever type, have in the past always seemed to go slower than those steered by men because we were weighed down by more diverse responsibilities. But going slower in retirement is then not as great a change of pace as it is for many men, who have steamed ahead all their lives. Most women therefore have an easier time with the transition.

RE-TIREMENT

You have another option: You can re-*tire*.

To re-tire. Wouldn't you think that meant to put new tires on an old car, to make it run like new and still have a useful life? Well, here you are, at 62 or 65 or 70, about to be retired. You have been rolling along (or bumping along) these many years and your tires are getting a bit worn, for the road was not always smoothly paved, and the directions not clearly marked. In some places the road was not paved at all—no one had been that way before to mark the road and make a track. However, *your* tires have left a mark behind them; they have flattened the road a bit, packed down the dirt, made a furrow to indicate direction. It will be easier for those coming behind you.

You may have enough background to consider retooling for career counseling or vocational guidance; becoming a consultant to organizations in your particular area of expertise, or possibly on male/female dynamics at work; going on speaking engagements, teaching evening courses, and writing articles. After everything you have gone through, with everything you have learned, you are a valuable and still all-too-rare role model. It would be wonderful if you could make yourself available to younger women about to start on their journey, on their *Paths to Power.*

Carl Jung postulated that the second half of life is compensatory for the first: Men reclaim their denied femininity and women reposses their masculinity. In fact, women may reach their most integrated point in later life, which further qualifies us for re-tirement:

> As women become more assertive, more demanding, more self-confident in their later years, men tend to become gentler and more emotional. There may be a conflict as men become less involved in their work and seek more personal, meaningful relationships, and women switch from investing in relationships to investing more in their careers. As the men become more inner-directed, the women become outward-oriented.[1]

You get re-tired to begin a new life. Of course, you must still assert yourself and make demands in order to pursue your goals. Do no sit and wait for others to make the overtures, just as women have in the past waited to be discovered, to be hired, to be promoted, to be loved. It is up to you to take the initiative, to start something, to open up opportunities for yourself, to establish new social and business networks.

Besides consulting, counseling, and teaching, another option for a re-tired woman is entrepreneurship. It is a risky undertaking (between 50 and 70 percent of all enterprises fail in the first five years) and one that few women can start, since women (especially older women) are still considered to be poor credit risks for venture capital.* However, as more and more women break into top management and prove themselves to be qualified business innovators, that restriction will ease.

What traits characterize an entrepreneur? High self-esteem, a willingness to take risks, independence, competitiveness, a strong need to achieve, energy, and perseverance predominate. Note that a good manager does not necessarily make a good entrepreneur, since their motivational needs (power versus achievement) are not the same. Do these characteristics describe you?

If starting your own business seems exciting, remember that it is an all-encompassing enterprise, requiring the full commitment of your time and energy. If you do not want to work long hours and worry through long days, entrepreneurship may not be for you. However, you can still work at something less time-consuming; working two or three days a week, or just in the afternoons, or only over weekends can be rewarding. Besides, although your spirit, energy, determination, and skills may not have diminished, an older body *does* have its limitations, even when in the best of health. It's important to recognize whatever limitations you have and to avoid straining to go beyond them.

Late starters

There are more older women in our society today than ever before. That group now numbers more than 18 million of a total of about 215 million. Some researchers predict that by the year 2000, two out of every three elderly (defined as age 60 or older)

---

*Venture capital is money used exclusively to explore and exploit investment opportunities in new or relatively untried business areas.

will be a woman. In the last twenty-five years, the number of older women who are working doubled.

Many of these women are recently divorced or widowed and must work simply to survive. These women are called displaced homemakers. Others are still married to an income-producing husband and work to earn extra money, to fill a void, or to return to a long-interrupted career. These are the late-entry women.

Displaced homemakers are frequently unprepared and undereducated for entry into the work force. Their training and skills are often outdated—sometimes as much as twenty or thirty years have passed since their last educational or employment experience. In addition, these women usually lack the self-confidence to overcome the age and sex restrictions still common in the work environment. To make a bad situation worse, they generally must accept whatever work is offered for financial considerations instead of getting the educational credentials required for better positions.

Late-entry women may have an easier time, simply because their income is not the only income. Many work to establish a new identity once their child-rearing years have passed. More and more older women are returning to school or starting their own businesses, many with financial help from family members. Still, they encounter the same problems experienced by displaced homemakers.

For those women who, at a late age, are either entering the workforce for the first time or returning to the business world after a long interruption, for whatever reason, I want to reiterate that, hard as it may seem, it *is* possible to succeed. The skills developed during a lifetime at the traditional feminine tasks can be translated into the competencies required by organizations. First, *you* must recognize this; second, you must make *prospective employers* aware of it. Please read Chapter 2 for a more complete discussion of this theme.

LOOKING FORWARD

I suspect that the discrimination experienced by older women may ease somewhat in the future. Because of declining birth rates, the elderly comprise an ever larger proportion of our population, a fact not unnoticed by the advertising industry. Their slogan, "You're not getting older, you're getting better," is, I hope, a harbinger of a new attitude toward aging. Since women outlive men by an average of seven years, the majority of the elderly who will profit from this change are women.

Despite what the rest of society thinks, growing old is not the end of living. Many men and women have aged with dignity and grace to discover another—sometimes richer—phase of life. Dame Agatha Christie wrote eloquently about this period:

> I have enjoyed greatly the second blooming that comes when you finish the life of the emotions and of personal relations; and suddenly find—at the age of 50, say—that a whole new life has opened before you, filled with things you can think about, study, or read about. You find that you like going to picture exhibitions, concerts and the opera, with the same enthusiasm as when you went at 20 or 25. For a period, your personal life has absorbed all your energies, but now you are free again to look around you. You can enjoy

leisure; you can enjoy *things*. You are still young enough to enjoy going to foreign places, though you can't perhaps put up with living quite as rough as you used to. It is as if a fresh sap of ideas and thoughts was rising in you. With it, of course, goes the penalty of increasing old age. . . . But one's thankfulness for the gift of life is, I think, stronger and more vital during those years than it ever has been before.[2]

Our later years can and should be a fulfillment of our lifetime. With more tolerant attitudes this will be easier to achieve. We have an obligation to help these attitudes emerge, for ourselves and for those who follow. An active, involved older woman is a wonderful role model for all people of all ages.

We must begin to look with new eyes, to hear with new ears, to think thoughts that until just a generation ago were unthinkable. We are in the process of attaining a new status—a new "place"—in the world, fostered by expanded freedom and a deeper understanding of what women are, want, and can do. We are learning that successful self-realization does *not* mean giving up our qualities of womanhood, it means adding on to them. By celebrating our knowing as well as our knowledge, we can be feminine *and* powerful.

You and I have journeyed together on the *Paths to Power*. I have shared my knowing and my knowledge with you; now it is your turn to share yours with other women. Together, we make an endless chain of powerful, empowering women.

NOTES

1. David Gutmann, "Life's Compensation for the Battle of the Sexes," *International Herald Tribune,* February 4, 1974.
2. Agatha Christie, *An Autobiography* (New York: Ballantine/Random House, 1977), pp. 630–631.

SELECTED READINGS

Ginzberg, Eli. *Career Guidance: Who Needs It, Who Provides It, Who Can Improve on It.* New York: McGraw-Hill, 1971.

Jaslow, Philip. "Employment, Retirement and Morale Among Older Women." *Journal of Gerontology* 31–2(1976).

Pospisil, Vivian. "Women in Management: Pattern for Change." *Harvard Business Review,* July–August 1971.

Starting and Managing a Small Business of Your Own. Washington, D.C.: U.S. Small Business Administration.

Timpano, Doris M., and Knight, Louise W. "Sex Discrimination in the Selection of School District Administrators: What Can Be Done?" Washington, D.C.: U.S. Department of Health, Education and Welfare, 1976.

"Women in the Labor Force: The Early Years, the Middle Years, the Later Years." *Monthly Labor Review,* November 1975.

Appendixes

A: RESUME PREPARATION

B: COVER LETTERS

C: THE POWER DIMENSIONS OF YOUR JOB

D: MULTICOMMITMENT

E: THEORIES THAT HAVE INFLUENCED MANAGEMENT THOUGHT

F: SOCIAL SECURITY PENSIONS

Appendix A: Résumé Preparation*

WHAT DOES A RÉSUMÉ LOOK LIKE?

The eye always travels to the upper third of a résumé first, so the most important factors—whether educational background or employment experience—should be placed toward the beginning. However, there is nothing immutable about the sequence of résumé items.

The chronological résumé

If you are graduating from college and have had little work experience, you will want to write a *chronological* résumé. This will include:

1. Your name and campus address as well as your personal (home) address.

2. Professional or career objectives. This section states what you are looking for in a position or training program. This category can be a "Catch 22," however: If you are too specific you may limit your job options, while an objective that is too general is worth little. If you apply for a variety of jobs, you may want to prepare two résumés, one that identifies a specific job or objective, and one that omits this heading and addresses career intentions in a cover letter.

3. Educational background.

4. Employment experience (including volunteer work).

5. Your personal background. Include this material only if it relates to the position you seek.

6. Your interests (again, as they relate to the position).

7. Personal information. This section should be included only if it will enhance your chances of being hired.

You are not required to provide marital status, or information about your family.

*I gratefully acknowledge the help of Barbara Babkirk, Associate Director of Career Planning and Placement of the University of New Hampshire.

SAMPLE CHRONOLOGICAL RÉSUMÉ Joan is 21. She has recently completed a Bachelor's Degree in Business Administration and is seeking an entry-level position in bank management. Although her work experience has not been extensive, her résumé highlights the skills she has developed through extracurricular activities and academic training. One source of Joan's information: the College Career Placement and Planning Office.

The recruiting schedule might appear as follows:

Employer and Date of Visit	Openings and Requirements	Requests Interviews With:
Sure Save Banks February 18	Bank Management Trainees	B.S. Candidates in Administration and Economics

Joan's résumé might look like this:

JOAN E. ADAMS
18 Commonwealth Avenue
Boston, Massachusetts 02115
(617) 536-1210

EDUCATION

College of Massachusetts	Boston, Mass.
Bachelor of Science	May 1980
Major: Business Admin.	Minor: Computer Science

Extracurricular Activities:

Member, Inter-Dorm Council; Business Manager, Campus *Gazette*; Treasurer, Computer Club; Member of tennis team.

EXPERIENCE

Business Manager, Campus Gazette, College of Massachusetts, Boston, Massachusetts. Responsible for: Managing and directing all accounting functions, scheduling office operations, hiring and supervising student staff.

Intern, Bank of Amherst, Amherst, Massachusetts. Responsible for: Assisting loan officer, reviewing mortgage and loan applications, preparing related documents, providing customer service in completing loan applications.

WORK HISTORY

1979 (summer)	Y.M.C.A. Camp Fun, North Adams, Massachusetts Camp Counselor, Tennis Instructor
1976–1978 (summers & vacations)	Barnard's Fine Apparel, Boston, Massachusetts Salesperson

PROFESSIONAL SKILLS

Organized   Efficient   Accurate
Able to delegate authority responsibly
Possess good verbal and written communication skills

REFERENCES

Will be furnished upon request.

JOAN E. ADAMS
18 Commonwealth Avenue
Boston, Massachusetts 02115
(617) 536-1210

EDUCATION

College of Massachusetts   Boston, Mass.
Bachelor of Science   May 1980
Major: Business Administration   Minor: Computer Science

Extracurricular Activities ◄──────── **College activities indicate leadership and business experiences that relate to future job goals**

Member, Inter-Dorm Council: Business Manager, Campus *Gazette*; Treasurer, Computer Club; Member of tennis team.

Past experiences ──────────► EXPERIENCE
are highlighted in terms
of tasks performed.
Shows demonstrated skill
areas

Business Manager, Campus Gazette, College of Massachusetts, Boston, Massachusetts. Responsible for: Managing & directing all accounting functions, scheduling office operations, hiring and supervising student staff.

Intern, Bank of Amherst, Amherst, Massachusetts Responsibilities included assisting loan officer, reviewing mortgage and loan applications, preparing related documents and providing customer service in completing loan applications.

WORK HISTORY ─────➤ **Employers may wish to see a work history even if it does not relate to a specific job. Indicate employer, address, position, and dates (if you do not have gaps in your employment)**

1979   Y.M.C.A. Camp Fun, N. Adams, Massachusetts
(summer)   Camp Counselor, Tennis Instructor

1976–1978   Barnard's Fine Apparel, Boston, Massachusetts
(summers & Salesperson
vacations)

Qualifications or ◄──────── PROFESSIONAL SKILLS
**skills statement
allows you to**   Organized   Efficient   Accurate
summarize abili-   Able to delegate authority responsibly
ties that are not   Possess good verbal & written communication skills
**obvious from
reading expe-
rience and work
history. Match-
ing your skills
with those re-
quired in pro-
spective jobs
might add
strength to your
résumé**   REFERENCES
                         Will be furnished upon request.

APPLYING THIS TO YOU   Look beyond your paid jobs for relevant experiences. By considering these extracurricular activities, determine the skills you have developed and used. Present these skills in relation to the qualifications of the job you seek. Supplement a brief work history with a statement of skills or qualifications.

The functional résumé

If your strongest selling points are neither your educational background nor your job experience, you should use a *functional* résumé, which stresses your range of skills and talents. The functional résumé allows you to focus on skills needed for any volunteer activities and responsibilities you might have undertaken, and is usually more appropriate for an older woman who has not worked for many years, or for the woman who has never worked at all.

SAMPLE FUNCTIONAL RÉSUMÉ   Beverly is 34 years old. She has three children, ages 12, 11, and 9. She attended college for two years before dropping out to marry and relocate at her husband's place of employment. Then, for two years, she worked as a legal secretary/administrative assistant. After the birth of her first child, Bev stayed at home for ten years taking care of her three children.

When her children reached school age, Bev re-entered the work force. Having enjoyed her experience as a legal secretary, Bev contacted law firms and legal agencies. She was offered and accepted a position as a receptionist with a legal rights advocacy organization.

At this point, Bev has worked for one year and feels she is capable of assuming more responsibility on the job. In order to pursue and test her interest in legal work, Bev has enrolled in an evening program in paralegal studies. She is currently considering an upward job change and will be distributing her résumé to appropriate agencies.

One source of Beverly's job information: *Newspaper Want Ad*

PARALEGAL ASSISTANT

Immediate vacancy. Qualifications: Experience and/or degree in political science or related area of study.
Send résumé to:
    Human Rights Coalition
    4 Emery Street
    Boston, Mass. 02001

Note: When a person's name is not indicated in an ad such as this, it is wise to contact the organization and ask for the name of the person to whom you should direct correspondence.

BEVERLY SMITH
60 WEST STREET
DURHAM, NEW HAMPSHIRE
03824
(603) 868-1712

Job Objective: Position as a Paralegal Assistant

Experience: -LEGAL-

- Researched legal issues concerning municipal zoning for political campaign
- Prepared legal papers and correspondence
- Reviewed law journals and other legal publications to identify pertinent cases
- Familiar with legal terminology and concepts

-ADMINISTRATIVE-

- Organized general office procedures
- Supervised clerical staff of three
- Scheduled meetings and appointments for staff of six
- Established and maintained new filing system
- Assessed client's needs and referred them appropriately
- Screened all phone inquiries

Work History

1967–1969   Seacoast Legal Assistance
                45 Pleasant Street
                Portsmouth, New Hampshire
                Receptionist

1979–1980	Noyes, Harmon & Smith
	14 South Street
	Durham, New Hampshire
	Administrative Assistant

Education:

1979 to present	University of New Hampshire
	Durham, New Hampshire
	Currently enrolled in Paralegal Studies certificate program.
1965–1966	Rivier College
	Nashua, New Hampshire
	Completed 2 years of undergraduate program
	major area of study: English
	G.P.A. 3.3/4.0

References: Will be provided upon request.

Name in right margin provides
easy access for employers
when flipping through files

BEVERLY SMITH
60 WEST STREET
DURHAM, NEW HAMPSHIRE
03824
(603) 868-7712

Job objective: Position as a Paralegal Assistant

Job objective is very specific
and sets the theme for the
résumé

Experience: LEGAL

- Researched legal issues concerning municipal zoning for political campaign
- Prepared legal papers and correspondence
- Reviewed law journals and other legal publications to identify pertinent cases
- Familiar with legal terminology and concepts

Experience identified by
related skill areas

ADMINISTRATIVE

- Organized general office procedures
- Supervised clerical staff of three
- Scheduled meetings and appointments for staff of six
- Established and maintained new filing system
- Assessed clients needs and referred them appropriately
- Screened all phone inquiries

Aims for concise yet precise
use of verbs that are very
descriptive of job tasks

Work History

1967–1969   Seacoast Legal Assistance
               45 Pleasant Street
               Portsmouth, New Hampshire
               Receptionist

**Provides info on previous
employers. Does not need to be
detailed**

1979–1980   Noyes, Harmon & Smith
               14 South Street
               Durham, New Hampshire
               Administrative Assistant

Education:

**Placed after *Experience*, which
might compensate for lack of a
degree**

1979 to present   University of New Hampshire
               Durham, New Hampshire
               Currently enrolled in Paralegal Studies certificate program.

**Continuing education shows
commitment and interest in
subject**

1965–1966   Rivier College
               Nashua, New Hampshire
               Complete 2 yrs. of undergraduate program
               major area of study: English
               G.P.A. 3.3/4.0

Honor grades may be included

References: Will be provided upon request.

**Not listing references allows
you to maintain control of who
is contacting whom on your
behalf**

APPLYING THIS TO YOU   The following factors relating to Bev's career development may increase her chances of securing meaningful work. You may consider them in your own career choices.

1. Identify an occupational area of interest (legal work).

2. Develop experience and skills in that field through volunteer and/or paid jobs.

3. Pursue opportunities to further refine skills and knowledge in the field (continuing education courses, in-service training programs).

4. Talk with persons who are in the field. Obtain information on opportunities for advancement.

5. Develop your résumé to clearly identify your skills and demonstrated abilities.

The untraditional résumé

One of your objectives in writing a résumé is to make it go to the top of the pile. How do you accomplish this? Make it different! There are many ways of doing this. If you think your competition is stiff and if your guess is that being "different" will not be seen as being *too* different, you might try to:

■ vary the size of the paper your résumé is on

■ vary the color of the paper

■ vary the texture of the paper

■ use an unusual type face

■ use a different colored ink

■ underline some items in red

■ use spacing attractively throughout

■ use a totally unique format, such as in the following two examples

SAMPLE UNTRADITIONAL RÉSUMÉ   Ellen is 53. Her youngest child has just begun college and she feels that she can now look for a job.

Ellen graduated from college 31 years ago. She married that summer and has since devoted her time and energy to her husband and children. She has also been extensively involved in community affairs. Because she has had no paid employment, Ellen thought it would be useless to write a résumé. However, after considering her volunteer experiences and listing her responsibilities, she was able to recognize and organize her skills into a functional résumé.

Sources of Ellen's job information:

1. Contacts from a previous informational research.

2. Interviews about career possibilities.

3. Newspaper articles and media notices about new programs.

Ellen Jackson   37 Strawberry Court   Portsmouth New Hampshire 03801   603 436-2980

Experience

1968 to present
Community Hospital
Coordinator of Volunteers (1979)

1977 to present
YWCA
Board of Directors
Executive Secretary (1979)

1949 to present
Alumni Secretary
University of New Hampshire
Class of 1949

1974–1979
Alumni Center
University of New Hampshire
Building Fund Committee Member

1960–1978
Heart Fund
Fund Drive Coordinator (1964–1968)

1957–1971
While children were growing up,
active in assuming leadership
roles in Boy Scouts, Cub Scouts,
4-H Clubs, Girls' Clubs, and
Parent Teacher Association.

Education

1966–1980
Continuing Education Courses:
University of New Hampshire
Northern Essex Community College
New Hampshire Vocational Technical
College

History (16 graduate credits)
Psychology
Interpersonal Dynamics
Grant Proposal Writing
Creative Writing
Gourmet Cooking
Modern Dance
Auto Maintenance and Repair

1949
University of New Hampshire

Bachelor of Arts
Major: History
Minor: French
Spent junior year at the
University of Dijon, France

References will be furnished upon request

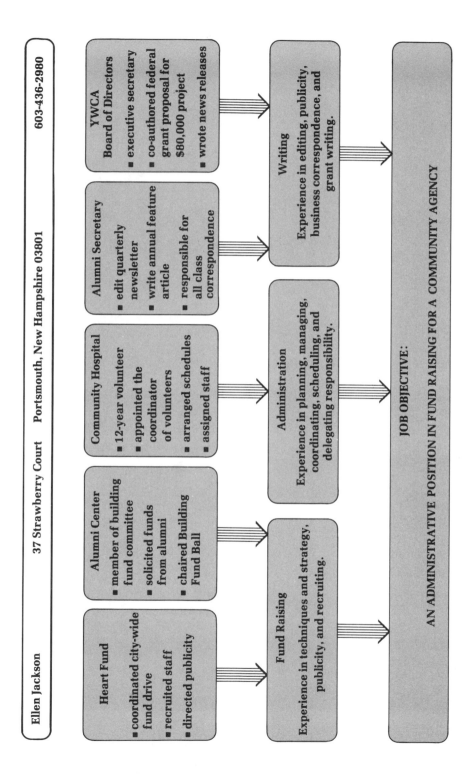

Ellen Jackson   37 Strawberry Court   Portsmouth, New Hampshire 03801   603-436-2980

Heart Fund
- coordinated city-wide fund drive
- recruited staff
- directed publicity

Alumni Center
- member of building fund committee
- solicited funds from alumni
- chaired Building Fund Ball

Community Hospital
- 12-year volunteer
- appointed the coordinator of volunteers
- arranged schedules
- assigned staff

Alumni Secretary
- edit quarterly newsletter
- write annual feature article
- responsible for all class correspondence

YWCA Board of Directors
- executive secretary
- co-authored federal grant proposal for $80,000 project
- wrote news releases

Fund Raising
Experience in techniques and strategy, publicity, and recruiting.

Administration
Experience in planning, managing, coordinating, scheduling, and delegating responsibility.

Writing
Experience in editing, publicity, business correspondence, and grant writing.

JOB OBJECTIVE:
AN ADMINISTRATIVE POSITION IN FUND RAISING FOR A COMMUNITY AGENCY

HIGHLIGHTING ELLEN'S RÉSUMÉ: Since Ellen's job target is in the creative field of fund raising, she uses an innovative résumé format. Although Ellen has never held a paying job, her skills and experiences are organized and presented as the appropriate "qualifications" for her job target. Experience is listed in both an innovative way and in a more traditional form. The traditional listing provides prospective employers with a more detailed history of work and educational experiences.

The format, although innovative, is logical and consistent throughout. The skills of fund raising, administration, and writing are identified as those most appropriate to Ellen's job target. Tasks and responsibilities are reported in specific terms that also relate to the job target: *coordinated, recruited, solicited, arranged, assigned,* etc. The cost of the grant is included since this substantiates Ellen's experience and responsibilities in the financial aspects of fund raising.

APPLYING THIS TO YOU: A functional résumé focuses on specific transferable skills. Although Ellen was never paid for her responsibilities, she did develop and successfully use a variety of skills. Her strategy in writing her résumé may be helpful for you. Recognize skills that you have developed and used, through volunteer activities, raising your family, or life experiences. Identify job targets. By doing some informational interviewing (where she asked the questions), Ellen found that many organizations have full-time professional fund raisers.

Develop your résumé in terms of your skills and your job targets. Use the appropriate jargon. Focus on skills related to your field of interest. The following stock data worksheet may help you in writing your résumé.

Personal Stock Data Worksheet

Personal Data Section
Full Name
Local Address   Tel. (   )

Permanent Address   Tel. (   )

Personal History

(Employment Experience)

Work History

(Be sure to include all full-time, part-time, unpaid/volunteer, summer, etc.)

Name & Address of Employer	Dates You	Position	Responsibilities
Contact Person for Reference	Worked	Title	or Duties

(This is for your own information, to be given out only if requested by employer)

Educational Background

Name & Address of Institution	Dates	Major and Minor	Degrees and
Contact Person for References	Attended	Concentration	Certifications

(This is for your own information, to be given out only if requested by employer)

Other Educational Experiences

(Include special seminars, workshops, conferences, mini-courses)

Publications

(Special Reports, Proposals, Written Projects, etc.)

Sororities, Associations, Professional Societies

Description or Name   Contact Person   Your Involvement   Dates

Honors, Awards, Special Recognitions

Travel (Location, Dates, Reason)

Special Interests, Hobbies, Avocations

Special Talents (Languages, Arts)

Military Service (Rank, Dates, Specific Duties)

Other Information

Résumé Critique Form*

Résumé of _____

Rate the résumé on the points shown below, scoring from a low of 1 to a high of 3 in each of the categories listed. Then score and compare your rating against the highest possible total score of 30. Write comments for each category receiving a score of less than 3.

Item	Score 1 2 3	How It Could Be Improved
1 Overall appearance. Do you want to read it?		
2 Layout. Does the résumé look professional, well typed and printed, with good margins, etc.? Do key sales points stand out?		
3 Length. Could the résumé tell the same story if it were shortened?		
4 Relevance. Has extraneous material been eliminated?		
5 Writing Style. Is it easy to get a picture of the applicant's qualifications?		
6 Action Orientation. Do sentences and paragraphs begin with action verbs?		
7 Specificity. Does résumé avoid generalities and focus on specific information about experience, projects, products, etc.?		
8 Accomplishments. Are applicant's accomplishments and problem-solving skills emphasized?		
9 Completeness. Is all important information included?		
10 Bottom Line. How well does the résumé accomplish its ultimate purpose of getting the employer to invite the applicant in for an interview?		

Rating Point Total __ (Out of a maximum of 30)

What are some other ways that you would suggest to improve this résumé?

_____

_____

_____

_____

*From _Résumé Preparatory Manual._ Reprinted with permission of Catalyst, 14 East 60 St. New York, NY 10022.

Appendix B: Cover Letters

Cover or application letters should always accompany your résumé when you're *not* distributing it in person. A well-written cover letter serves as an extension of your résumé. Since you will be composing an original letter for each employer, you can tailor the information to each particular job. Whenever possible, use the jargon of the job you are applying for.

The following points are important parts of the letter:

1. *Introduction:* Try to capture the employer's interest. Make that person want to read more about you. Mention your source of job information, an accomplishment that relates to the particular job opening, and state why you have selected this particular employer.

2. *Specifics:* Describe your interest in the position (or organization), referring to past experiences through which you have demonstrated this interest. Include any special skills or abilities that might further qualify you. Refer the employer to your résumé to complement this information.

3. *Closing:* Suggest a follow-up, preferably an interview. You might indicate that you will be calling to determine the status of your application—*or* you might conclude with a question,—asking the employer when you might hear about your inquiry. In either case, take responsibility for following up within a few weeks.

TIPS ON WRITING A COVER LETTER

1. Each letter of application should be an original typed letter. Duplicated letters are *not* appropriate.

2. Use simple, direct language and correct grammar.

3. Your letter should be flawlessly typed.

4. Keep the letter short and concise.

5. Use a standard business letter format with the body centered on the page and equal margins on the sides.

6. Use quality bond paper, 8 1/2″ × 11″, either white or in a color matched to your résumé.

7. Keep a copy of your correspondence for future reference.

8. Your letter should be an introduction to your résumé, not a review of it.

The following examples involve three women considering employment. They have very different backgrounds and skills. Consider their approaches to their job search.

[Date]

Ms. Frances Stoll
Personnel Manager
Sure Save Banks
6701 Burbank Place
Newton, Massachusetts   09000

Dear Ms. Stoll:

Having recently registered with the Career Planning and Placement Office at the College of Massachusetts, I was pleased to learn that Sure Save Banks will be recruiting on campus this month. I have scheduled an appointment with you on February 18, and I am eager to discuss my qualifications and background in light of employment opportunities with Sure Save.

For your review, I have enclosed a résumé that indicates my undergraduate program as well as experiences that have helped me combine course work with actual business settings.

It will be a pleasure to meet with you.

Very truly yours,

Joan E. Adams

[Date]

Ms. Alice M. Stevens
Executive Director
New Hampshire United Way
43 Pleasant Street
Concord, New Hampshire   03301

Dear Ms. Stevens:

First of all, I want to thank you again for the time you spent with me in the fall and for all of the information you gave me about career possibilities in fund raising. Since talking with you, I have investigated many other options for careers with community organizations and have determined that the appropriate choice for me is administration of a fund raising program. As you so correctly pointed out, this is a logical choice for me considering my interests and experience. Having made this decision, I am now actively seeking a position.

This brings me to my second point. It is my understanding through news and media reports that the New Hampshire United Way will be involved in a major fund raising campaign over the next two years with the goal of increasing funding sources by two hundred percent. I would like to be a part of that effort. To that end, I am enclosing a copy of my résumé for your review.

Although I realize that you may not have a staff opening presently available, I would appreciate the opportunity to talk with you about the possibility of my working with the United Way. I will telephone you early next week in hopes of arranging an appointment. I look forward to meeting with you soon.

Sincerely,

Ellen Jackson

[Date]

Paul Sacks, Director
Human Rights Coalition
4 Emery Street
Boston, Massachusetts   02001

Dear Mr. Sacks:

Attorney Susan Riley, my present employer at the New Hampshire Legal Assistance Office, suggested that I contact you regarding the current paralegal vacancy in your organization.

In March I will have completed the certificate program in Paralegal Studies at the University of New Hampshire at Durham. This course of study, coupled with my previous experience with legal procedures, has provided me with excellent opportunities to develop skills and knowledge in the field of paralegal assistance.

As my enclosed résumé indicates, my experience includes both administrative and legal responsibilities. Since my former work settings range from private legal practice to a state-subsidized legal advocacy organization, I believe that I am well qualified to deal with the broad variety of issues faced by the Human Rights Coalition.

I am looking forward to an opportunity to further discuss my qualifications with you. When might we arrange an interview?

Sincerely,

Beverly Smith

Follow-up

Whenever you interview with someone, in person or by phone, always ask for the names of other people you might talk with. And keep a file of the information you've collected during your search for a job: You never know when it will come in handy during some future job hunt. Finally, any interview or telephone conversation should be followed by a polite letter of thanks. Even a brief thank-you note will enable you to remain visible in the future employer's eyes.

Correspondence Checklist*

Letter of Application
1. Identify the position for which you are applying and how you learned of the firm and position.
2. Indicate why you are applying for this particular position.
3. Describe your main qualifications.
4. Refer the reader to the enclosed résumé.
5. Request the next step in the employment process—personal interview, an answer to your letter, etc.
6. Be sure to sign the letter.

Letter of Acknowledgment
1. Acknowledge receipt of offer.
2. Express your appreciation for the offer.
3. Notify the company of the date you expect to make your decision.

Letter of Inquiry of Application Status
1. Request status of application.
2. Recap history of your application.
3. State why you need clarification of status of application.

4. Include thanks for cooperation.

Letter Declining Offers
1. Decline offer.
2. Express your appreciation for the offer and the company's interest in you.

Letter Seeking Additional Information
1. Indicate interest in the company and its offer.
2. Ask for the information you need. Be specific!
3. Express your appreciation for the co-operation you receive.

Letter of Acceptance
1. Accept the offer.
2. Refer to offer letter or document.
3. Tell your travel plans and anticipated arrival date.
4. Express your appreciation and your pleasure at joining the company.

*Copyright 1977 *College Placement Annual.* Reproduced by permission. Prepared by Prof. James W. Souther, former Director of University Placement Services, University of Washington.

FINDING THE JOB THAT'S RIGHT FOR YOU

To find the right job, you need to determine what you can do best. Narrowing down a career choice may seem like a formidable task, but it's like solving a mystery: There are clues everywhere about what you can already do, what you want to do, and where the jobs are. And once you've been through the process, you've learned *how to find a job,* which is quite a feat, after all. The job-hunting procedure itself demands many skills that are probably transferable to the job you ultimately accept.

The more you know about various companies and the sort of jobs that are available, the easier it will be to set career objectives and market yourself in a particular field. One place to begin looking is *The Occupational Outlook Handbook*, which can be found in most local libraries. This book describes hundreds of job positions and supplies information on job demand, general working conditions, salary, and job qualifications. By matching your own skills with various job descriptions in *The Occupational Outlook Handbook*, you can identify those jobs that both interest you and suit your own capabilities. *Woman's Work*, a bi-monthly magazine, also contains helpful suggestions for this stage of the job search. The magazine offers descriptions of the latest national and regional trends in the job market.

Here's a list of additional sources that may help you in your search:

1. A helpful reference librarian.
   Invaluable—especially if you know how to ask the right questions. And most reference librarians are only too willing to help out. They'll also show you how the library works, and that knowledge is most useful in the job-hunt process.

2. *Finding Facts Fast: How to Find Out What You Want To Know Immediately*, by Allan Todd
   A great guide to beginners in the research game that contains lots of helpful suggestions.

   Find out the name of the local newspaper (local to the company you're interested in, that is) and ask the business editor your questions, or ask where you might go to find the answers. Often, newspapers have librarians and researchers who can be of help. *Editor and Publisher International Yearbook* lists the names of all local news publications in the United States and Canada.

   Consult your college alumni directory, which should have information on where your classmates are, what they're doing, etc. Maybe one of them works in the field you're interested in. This suggestion applies to any lists of people you may have access to—high school directories, clubs, sports teams, volunteer organizations, etc.

3. *Encyclopedia of Business Information Sources*, compiled by Paul Wasserman
   Provides the names of trade associations, periodicals, directories, almanacs and yearbooks, biographical sources, statistics sources, bibliographies, etc., of the particular industry you may be interested in.

4. Directories of Organizations Within Industries
   Many industries list their organizations—complete with addresses, a short description, and a contact person. For example, the market research profession has a "green book," which lists most market research houses in the country. The advertising profession has a "red book." Ask your reference librarian if the field you're interested in has such a directory.

5. *Standard and Poor's Register of Corporations, Directors, and Executives*
   Lists thousands of leading business firms, their products, number of employees, and the names and positions of key employees. It also contains geographical and product classification indexes.

6. *Funk and Scott Index of Corporations and Industries*
   Company information is presented alphabetically by company name. Includes listings of articles appearing in all major business, financial, and trade magazines, key newspapers, analytical reports of investment advisory services, and bank newsletters.

7. *Moody's Manuals*
   Provides general information on most publicly-owned companies, including the names of key personnel and financial status.

8. Business Periodical Index
   Good for finding current articles written about the company or field you are interested in.

9. *Wall Street Journal* and *New York Times* Indexes
   The *Wall Street Journal* and *New York Times* often highlight particular industries, companies, or types of positions. Perhaps what you're looking for has been analyzed in either—or both—of these papers. Find out by using their indexes.

10. Annual Reports
    These always contain financial information, and often describe equal opportunity efforts, new products, and anything that the company is proud of.

11. *Working Press of the Nation, Volume 5: Internal Publications Directory*
    A guide to internal publications of companies, which are useful for getting day-to-day flavors of the company you might be researching. (Used to be known as *Gebbie House Magazine Directory.*)

12. *Women and Work* Office of Information, Publications, and Reports, U.S. Department of Labor, Washington DC 20210

13. Equal Employment Opportunity Commission, Office of Research, 1800 G. St., NW, Washington DC 20506

14. *Equal Employment Opportunity Report: Job Patterns for Minorities and Women in Private Industry*, U.S. Equal Employment Opportunities Commission, Washington DC 20506

15. U.S. Women's Bureau, Washington DC 20210

16. *Women's Rights Almanac*, published by Harper & Row

17. American Association for Affirmative Action, Administration Building Room 210, Ball State University, Muncie, Indiana, 47306 (317) 285-5162.

One last word of advice: Don't throw any of your data away. In a few years, you may be looking for a more challenging, high-paying job, and all this information will come in handy.

Appendix C: The Power Dimensions of Your Job

BY THE STAFF OF
Goodmeasure, Cambridge, Massachusetts

Studying the power dimensions in your job is one of the more valuable exercises you can do to enhance the probability of being promoted. You can chart your *Paths to Power* by being not only good, but noticeably good (visibility), by taking on more responsibilities (autonomy), by developing subordinates and working with peers and bosses (relationships), and by pursuing company goals (relevance). It is usually more effective to do the action planning with a friend who can monitor you on your progress, since it is not easy to push yourself without support.

ANALYSIS FOR ACTION*

This questionnaire serves two functions. First, it helps you analyze the *current* aspects of your job that contribute to your effectiveness and to your prospects for developing a reputation for effectiveness. Second, it urges you to think of *future* actions that can develop those specific parts of your job known to contribute to job efficacy and organizational power. There are two parts to most of the questions: Part **a** asks about the way things now are; part **b** asks you to create action alternatives to improve the situation. Your responses to part **b** questions are an opportunity to *brainstorm* action possibilities for yourself. Even if you do not see a way to follow through on an alternative at present, list it anyway. The task is to develop a range of alternatives for yourself, with a minimum amount of censorship or a priori restrictions. In a later step you will be able to select from among these alternatives the most feasible ones to pursue. But right now it is important to be as creative and expansive as possible. Leave no question unanswered, even if the alternative you suggest is seemingly small or insignificant. (If you do not know the answer to a part **a** question, use part **b** to list ways you could learn the answer.)

Visibility: The extent to which your work is known in the organization.

1. a) With whom in the organization have I shared my ambitions, my career goals?

b) Who else could help me if they had more such information about me?

2. a) How is what I do in my job communicated to the larger organization (both informally and formally—e.g., meetings, conversations, reports)?

   b) How can I improve on this communication? What other forms are available to me?

3. a) Who are the influential people *within* my department who know about my work?

   b) What additional information about me could be useful to share with them?

4. a) Which of my job activities bring me into contact with people beyond my department or organization?

   b) How can these activities be expanded?

5. a) Who are the influential people *outside* of my department who know about my work?

   b) How could they learn more about me?

6. a) How often do I participate in committees, task forces, or other work groups that include people from across the organization?

   b) What other activities of this type can I become involved with?

7. a) What groups and professional associations outside of my employing organization do I belong to?

   b) How could increased participation in such groups benefit me?

Autonomy: The amount of discretion in your job

1. a) At present, what types of decisions do I make acting on my own authority (e.g., budget expenditures, hiring, staffing, etc)?

   b) In what other areas could I expand my discretionary authority? How could I get such authority?

2. a) What parts of my job allow me the opportunity to act on my own initiative, to demonstrate my creativity?

   b) How could I make these a more central part of my job activities?

3. a) Are there any new projects or activities in the organization or within my job that would give me the opportunity to create and develop something?

   b) How could I get involved with these projects or activities?

Relevance: The value of your job to pressing organizational issues

1. What are two to four crucial issues that my institution faces within the next three years?

2. a) Which aspects of my job play a role in addressing these issues?

   b) How can I further develop my job to become more involved with these issues?

3. **a)** What skills do I have that are the most value to the organization?

   b) How can I use these, or trade on them, to improve my job situation?

4. **a)** What other skills could I develop or improve on to make me more of an asset to the organization?

   b) How can I go about this?

Relationships: Supports and alliances on the job

1. **a)** How often am I in contact with my peers?

   b) How could I benefit more from these interactions?

2. **a)** How often am I in contact with women in roles similar to mine?

   b) How could I benefit more from these interactions?

3. **a)** Among my subordinates, who are the most promising?

   b) What things could I do to help these people develop and advance in their careers?

4. **a)** At present, what responsibilities do I share with my subordinates?

   b) What other of my responsibilities could be shared with them?

5. **a)** Which senior person(s) could best help me do my job more effectively?

   b) What can I do to develop this relationship(s)?

   c) How can I be of help to this person in his/her job?

6. **a)** Who is the most senior person with influence who has shown an interest in my career?

   b) In what ways can I make my skills and ambitions better known to this person?

   c) How can I be of help to this person in his/her job?

7. **a)** What contacts, both in and outside of the organization, do I have that are of most value to the organization?

   b) How can I use these, trade on them, to improve my job situation?

8. What contacts could I develop that would make me more of an asset to the organization?

ACTION PLANNING

From the many alternatives you have suggested to yourself, select three to five that, given your particular job situation, would do the most to increase your job effectiveness and/or career prospects. Let these be your goals. Write down each goal in the space provided. Then, for each goal, list one or two specific actions you can *immediately* do to begin achieving that goal.

GOAL: _____

_____

1st ACTION(S): _____

_____

_____

GOAL: _____

_____

1st ACTION(S): _____

_____

_____

GOAL: _____

_____

1st ACTION(s): _____

_____

_____

GOAL: _____

_____

1st ACTION(S): _____

_____

_____

INTERVIEWING KEY ADMINISTRATORS

One block that many women experience is the lack of opportunity to talk with upper-level people in order to find out how they obtained their positions. Interviewing key administrators provides a chance to be visible and possibly to exchange some ideas. The following interview guide provides a way to take full advantage of such an opportunity. The guide points out the topics that are critical in any career path, showing your initiative and imagination. Most women have found that key administrators are flattered to be interviewed and glad to talk about themselves. Making such a study demonstrates your ambition and willingness to take risks and to learn. Be careful not to do this on company time; instead, invite the manager out for lunch or for a drink after work. You should learn a great deal from the answers you receive.

Interview Guide*

Try this introduction:

> I am making a study of career paths, and I am speaking to a few key people here about their own career patterns and preparation. Our conversation should take no more than 30–45 minutes. Everything you say will be held confidential.
>
> I very much appreciate your taking the time to contribute your experiences and knowledge, and I hope you will find the chance to discuss these matters as interesting and valuable as I do.

Background

1. *Vital statistics:*
   Name of person interviewed
   Title and length of time in that position
   Brief description of functions, duties, and responsibilities (Does the job include hiring or participating in hiring decisions?)
   Sex
   Approximate age (under 30, 30–34, 35–39, 40–49, 50–59, 60–64, 65 +)
   Highest degree earned, institution, date, and field

2. *Sources of satisfactions:* Thinking back over the last year or so, what would you say have been the sources of greatest satisfaction to you in your job—perhaps some things you've accomplished, or perhaps just the ongoing things that you find satisfying?

3. *Headaches:* What have been the biggest headaches or sources of trouble?

4. *Skills:* a. What would you now identify as the principal skills that someone in your position needs to do the job effectively? (These can include things you need to know, things you need to be able to do.)
   b. Are these skills the *same* or *different* from what they might have been a few years ago (i.e., have job demands been changing)?
   c. Do you think different kinds of *people* are typically being put into a job such as yours today—that is, people with different kinds of training, from different fields, or with different kinds of experience? Are the credentials to get into the job changing? Do you think they *should* be?

5. *Career history:* a. What has been your own background? Could you please identify the jobs you have held, beginning with the first after school? (Title, institution, and approximate dates.)
   b. How typical do you think your history is of the way people generally get into a job such as yours?

6. *Career influences:* What were some of the key events or people that positively affected your career? (Go into as much detail as you like.)

7. *Training:* a. Did you ever participate in a formal training program? If yes, what, where, when?
   b. Were you supported by an institution, or did you pay your own way?

8. *Learning the job.* a. How would you say you learned to do your present job? For example, how much of it came through:

*Reproduced by permission of Goodmeasure, Cambridge, Massachusetts.

prior experience in other jobs (which)?
formal education or training?
being supervised and taught by someone (who)?
watching other people (who)?
on your own, on the job?

b. Do you think that the way you learned makes sense for other people, or would it be better in general to encourage another way of learning the ropes for a job such as yours?

"Now I'd like to ask a few questions about our institution's policies, as you see them."

9. *Growth:* a. How would you characterize our growth pattern over the last few years and into the next few—as *expanding, stable,* or *contracting?*
b. What effect has this had, or might this have, on what we look for in administration, and on the prospects for people already in administration?

10. *Policy changes:* a. Have there been any major policy changes over the past few years? Changes in "management" policies?
job criteria?
career paths?
affirmative action?
methods of appraising performance?
methods of recruiting administrators?
b. Do you anticipate any such changes in the next few years?

11. *Hiring patterns:* a. Have we at this institution tended, in the past, to promote internally, or have we tended to hire from other organizations (Please try to estimate the proportion of insiders versus external hires.)
b. Is this pattern changing?
c. In general, across administrative positions, are there signs of changes in the types of experience valued in recent job candidates (examples)?

12. *Women:* a Who are the principal women administrators (names, titles)?
b. What do you know of their backgrounds—e.g., prior jobs, institutions?
c. How do we hear of and/or recruit women candidates?
d. Is this different from the procedures for male candidates?
e. What do you think should be done to get more qualified women into higher positions?

Appendix D: Multicommitment

THE SEVEN-ROLES EXERCISE

1. Make a list of all your commitments, using the checklist below ("In my role as a").

2. Some roles are performed on a once-a-week basis. Calculate a daily average by dividing total hours worked per week by seven.

3. Make a note of your thoughts and feelings as you look at how you spend an average day.

4. a. Which roles provide you with satisfaction?
   Do you spend enough time in them?

   b. Which roles are unsatisfactory?
   Do you spend too much time in them?

   c. What will you change and how exactly will you do it?
   By when will the change be accomplished? Put this date on your calendar.

5. (*Optional*) Do the following exercise again, with your partner, asking him to substitute the words husband, father, son, and man.

In my role as a:

1. a. *worker*, I spend____hours a day at work (including commuting time).

   b. *worker*, I spend ____hours a day on the work I bring home.

2. *wife*, I spend   ____hours a day with my husband alone.

3. a. *mother*, I spend____hours a day with my children alone (including chauffeuring).

   b. *wife and mother*,
   I spend   ____hours a day with my family all together.

4. a. *homemaker*,
   I spend   ____hours a day doing housework* inside the house.

   b. *homemaker*,
   I spend   ____hours a day doing errands outside the house.

*Housework can be divided into cleaning, doing laundry, marketing and shopping, cooking, car and yard care, mending and repairing, paying bills and doing social correspondence, and telephoning about home- and child-related matters.

5. *daughter*, I spend ____hours a day with my parents.*

6. *friend*, I spend ____hours a day with my friends.*

7. a. *woman*, I spend____hours a day grooming myself.*

    b. caretaker of my
       body, I spend ____hours a day eating.
              ____hours a day sleeping.
              ____hours a day exercising (including walking).

      *Include telephone conversations.

My Average Day*

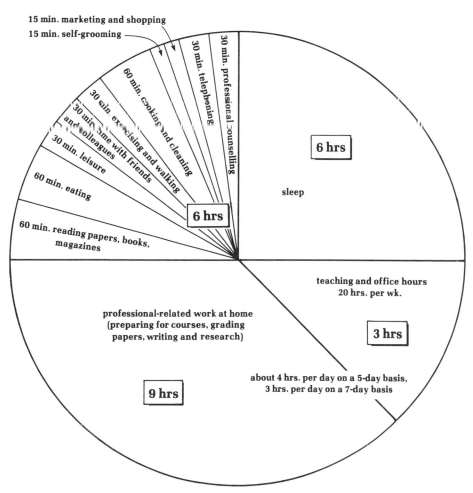

15 min. marketing and shopping
15 min. self-grooming
30 min. cooking and cleaning
30 min. exercising and walking
30 min. time with friends and colleagues
30 min. leisure
60 min. eating
60 min. reading papers, books, magazines
30 min. telephoning
30 min. professional counselling
6 hrs sleep
6 hrs
teaching and office hours 20 hrs. per wk.
3 hrs
professional-related work at home (preparing for courses, grading papers, writing and research)
9 hrs
about 4 hrs. per day on a 5-day basis, 3 hrs. per day on a 7-day basis

    c. caretaker of my
        mind, I spend ___hours a day reading or studying.
                    ___hours a day in hobbies.

    d. caretaker of my
        spirit, I spend ___hours a day alone, meditating, or doing nothing.

       TOTAL: ___hours a day

As you can see, I work too hard, don't exercise enough and have no time for myself! Here is one for you to fill out.

Your Average Day

Some people prefer doing it for the week, which would be a total of 168 hours. Others prefer dividing it into the five-day work-week with a separate one for weekends. You may want to shade all areas that deplete you of energy in contrast to the activities that energize you.

CAN YOU COPE?

If you have both career and family, you may encounter the following situations. How will you cope? There are no "shoulds." Whichever way you answer, there is a cost and a benefit. The questionnaire is to help you become aware of the possible dilemmas. Most items have no "solution"; instead, they need a compromise, which may not be completely satisfactory but is acceptable. The question is, how will you compromise and what can you negotiate in advance? The more thought you give to these, the less chance you will be caught off guard and unprepared.

1. You're about to graduate from college and are offered a great job. Your fiancé has an equally great opportunity elsewhere. What do you do?
   a. move with him
   b. take the job and postpone the wedding plans
   c. ask him to relocate
   d. take the job and cancel the wedding plans
   e. try an alternative living arrangement for a while (weekend marriage).

2. You are just married. Would you
   a. keep your separate banking and charge accounts?
   b. have separate accounts plus one joint fund contributed to equally by both?
   c. pool all the resources and operate out of one pocketbook?

3. Saturday is finally here—how do you spend it?
   a. grocery shopping
   b. tennis match and lunch with your husband
   c. at the zoo with the family
   d. attend a women's awareness lecture
   e. clean the house

4. You and your husband have had long-standing plans to go out to dinner Thursday night with friends you haven't seen in a month. On Monday, your boss tells you that there is a very important meeting Thursday night. Would you:
   a. forget the meeting and go out to dinner?
   b. cancel or postpone the plans with your friends and attend the meeting?
   c. attend the meeting while your husband goes out to dinner?

5. It's a weekday morning and everyone is late: You're not going to be at work on time, the children have missed the school bus, the phone is ringing, breakfast is cooking, and everyone is complaining. You:
   a. sit down and yell
   b. ask everyone to calm down and delegate the work that needs to be done

   c. leave for work, and leave the family to cope without you, which means your husband must take over

6. Regarding the mother, daughter, wife, and friend roles you play, at times you will be needed by several significant people at one time. For example, on the same day:

     your lawyer husband has just lost a case that was really important to his career and wants to talk to you about it;

     your daughter has just been rejected by a long-time boyfriend and needs your support;

     your college friend has called with the news that she is getting a divorce;

     you are in the middle of a vital project for your company.

     How would you divide your time?

7. You just received the raise that you've been waiting for, but last week your husband's company was subject to a big budget cut, and a good part of his salary went with it. Would you:

   a. rush in shouting the good news

   b. wait a few weeks to tell him

   c. secretly open a new account and put the extra money there, thinking that maybe at Christmas you can use it to get both of you something really special

   d. sit down and discuss the benefits to both of you

TWO-CAREER COUPLES

If you intend to *become* a two-career couple, I suggest that you ask yourself and your partner the following questions. Again, there are no "correct" answers. This is an instrument to help you identify possible future issues and problem areas. What is important is *talking together* about expectations from each other, personal dreams and hopes, concerns about the relationship, and the responsibilities each is willing or unwilling to assume. Attitudes and beliefs inculcated by early socialization should be reexamined in view of your new life tasks.

Career issues

1. What do you want out of life for yourself?

2. What is your current job standing? What is his?

3. What are your short-term goals in terms of employment (status, promotion, etc.)? What are your long-term goals? What are his? Five years from now where would you like to be working? Can you see yourself in this position if married to your partner?

4. How does your partner view your career?

   a. necessary for monetary reasons

   b. something to make you happy

   c. a part of your essential development as a person

   d. an interference with your being a good wife and mother or his being a good husband and father

5. What will have first priority, your marriage vows or your career goals?

6. Whose job has more priority in terms of relocation if the need arises? Upon what criteria will this be based?

7. Will you be willing to separate for extended periods of time if the job demands it?

8. Will you be willing to relocate if your partner gets a more satisfying job? Will you take a less desirable job (lower status or less pay) if your partner's job so demands?

9. Will you be willing to take a less desirable job while waiting for your partner to complete his education?

10. What will you do for vacation if your leaves fall at different times?

11. Are the career goals of you and your mate compatible?

Money and support issues

1. Who will be likely to make more money? Will the other resent this situation?

2. Will you want separate or joint accounts?

3. Out of whose salary will the daily living expenses (rent, car payments, insurance, phone, food, clothing, etc.) come?

4. Who will support whom, or will you share this role between you?

5. Whose responsibility is child support?

6. Is money a motivator for you? for him?

7. Would you choose a job because of the pay even if the location is not that desirable or conducive to the other's job?

Child care issues: Am I parent material?

1. Will both of you want a family? What size?

2. Will a family fit your life-style?

3. In what way would a child hamper your personal career development?

4. What effect would your dual-career marriage have on your child's development?

5. How will you determine who is responsible for the care of an infant?

6. Will one partner take time off from work or will you hire child care after the baby is born?

7. Does the area in which you live have an adequate educational (day care, pre-school, grade school, after-school program, high school) system?

8. How will you deal with an inadequate school system?

9. If the school system or town doesn't have day care, who will care for your pre-schooler?

10. Once in school, who will care for your child after school (3 to 5 p.m.)

11. Who will have the responsibility of arranging for a sitter for an evening out?

Division of labor issues

1. How will you work the responsibilities of keeping up your home?

2. Who will prepare the meals, purchase food, do the laundry, clean house, mow the lawn, clean the car, etc.?

3. If you choose to get a housekeeper, how and by whom will the payments be made?

Miscellaneous issues

1. In what type of setting do you wish to live?

2. If both of you work in the same field, will you view your partner as a colleague or as a competitor?

3. Will a partner's promotion cause admiration or resentment on your part?

ALTERNATIVE WORK SCHEDULES

There are five different categories of scheduling: 1. *staggered* hours; 2. the *compressed* work week; 3. *permanent part-time*; 4. *job sharing*; 5. *flexible* working hours. The statistics are evidence of their popularity. The latest estimates point to one million workers in compressed work week arrangements and ten million on flexible working hours. How do these alternative schedules work and how can women take advantage of them?

Staggered hours

According to this work schedule, groups of employees are assigned staggered arrival times. For instance, some might come to work at 6:00 a.m., some at 6:15, some at 6:30, and so on up until, let us say, 9:30. Consequently, the stopping times of the workers will be staggered, too. Thus, in this model, if the first of the workers arrived at 6:00 they will be the first group to go off, at 2:00. If a prospective employer speaks to you of staggered hours, be sure to find out if you have an input into what these hours will be for you, and try to come and go outside of heavy traffic hours.

Compressed work week

This arrangement permits organizations to compress the forty-hour week into less than five working days. The compressed work week is not a good solution for mothers of young children, because they are gone long hours during the day. It is, however, an excellent solution for a dual-career couple, who can spend a larger amount of time together at the end of the week, or for the husband and wife who are separated by some distance during the week. This permits them to be together for three consecutive days, provided both are on such a schedule. Some disadvantages of the compressed work week for a woman may be having to travel after dark or getting home too late in the evenings to participate in any kind of meaningful leisure activities.

Part-time employment

Part-time work can be on a part-day, part-week, or part-monthly basis. It can be permanently part-time, or it can be part-time for the individual during a particular phase of life. For example, you may decide to work part-time while you are raising young children, or while someone is sick at home that you need to take care of, or while you are going to school. The advantages of part-time work are evident: You can easily fulfill your other life commitments. The disadvantages, however, can be serious: You earn less money and, in addition, you are frequently seen as less committed to your career and are therefore less apt to be promoted or given larger responsibilities. Unless the job calls for part-time work, asking for part-time when the job specifies full-time may prevent you from landing the job. But in that case, there is still another option.

Job-sharing

This model involves two employees who are responsible for what would otherwise be the work of one full-time employee. They arrange their schedules to suit their needs as well as those of their employer, setting it up so that there is always one employee on the job, either on a half-day basis, a Monday-Wednesday-Friday and Tuesday-Thursday basis, or whatever other arrangement seems to fit all parties concerned. The job-sharers split both salary and fringe benefits. Job-sharing can be performed by two people who did not previously know each other, or by two people who come together to interview for one job. It can also be performed by a husband-and-wife team. Job sharing is not limited to two people splitting one full-time job. Two people can work three-quarter time for a job-and-a-half. In July, 1977, the Senate approved a bill put through by California's John Tunney, requiring that 10 percent of all federal job spaces be sharable within a five-year period.

Flexible working hours*

Flexible working hours can work within the eight-hour day, the forty-hour week, or the month. Within an eight-hour day, a person has the flexibility to arrive and leave at her own discretion or at a negotiated time with her colleagues and bosses, as long as she works eight hours. If it is within the forty-hour week, the flexibility would expand to working shorter hours on some days and longer hours on others, as long as the forty hours were completed by the end of the week. The month-long schedule gives an even wider latitude: Employees arrange their schedules according to their own needs and those of the organization, which could vary widely not only from day to day but from week to week.

The advantage of flexible working hours is the autonomy given individuals in controlling their own lives. Tardiness is reduced because it ceases to exist. When you arrive late, you just work later. Absenteeism is reduced because people can fulfill their other obligations without violating the normal working hours of their company. And finally, traffic congestion is reduced because of the different possibilities of arrival and departure time.

In 1978, several government agencies began experimenting with flexible working hours, waiving the overtime payments required up till now.

Fully 15 percent of the working population in America today are in some kind of variable work schedule. It has become legitimate for you to ask your company if it has a work schedule different from the normal 9:00 to 5:00. Alternative schedules work not only to the advantage of working women, but also of working men. For all people who have responsibilities other than those at work—for all people who want to pursue hobbies, leisure activities, or time with another—these variable schedules allow balance in their lives. As individuals gain more control over their time they will not only gain in autonomy, but in self-respect. We are living in an exciting time of transition. Alternative work schedules are part of that transition and part of the challenge of rearranging our personal and professional lifes to meet a broader range of human needs.

*Allan Cohen and Herman Gadon, *Alternative Work Schedules: Integrating Individual and Organizational Needs* (Reading, MA: Addison-Wesley, 1978).

Appendix E: Theories that have Influenced Management Thought

Why are management theories important? Why should we know such names as McGregor, Likert, Maslow, and Blake and Mouton? First, knowledge is one of the *Paths to Power*, and the conceptual base of your profession is the foundation of that knowledge. Second, the usefulness of theories lies in the increased possibility of predicting outcomes. If you can accurately predict an event, your planning strategies will be more effective and your reputation as a good manager more assured.

In any management training workshop or class, the people listed above, among others, are frequently mentioned as having substantially influenced the practice of management. The great majority of these theorists are men and most of their research is based on male samples, to be used by male managers. Since many women don't "fit" into the male-tested patterns, women are often seen as deficient. For example, theories that presuppose a specific personality profile for success can become deficiency models for women instead of permitting or promoting growth, understanding, and development; others attribute the same needs to men and women. Knowing the theories can help you place your organization and its workers and managers into a conceptual framework that will, in turn, enable you to see to what extent (if any) discrimination directly results from the application of a particular management theory.

I urge you to know what these people have written so that you can better understand the male assumptions and expectations. Whether you are going to be a boss or a subordinate, even a beginning understanding of some of the better known theories of management will increase your knowledge, your awareness, your credibility, and your self-confidence. Besides, just knowing the jargon of your profession may act as an admission ticket to the inner circle of leadership.

Management concepts fall into four main categories: what motivates people, especially workers; what managers do, which is descriptive; how they do it, which deals with leadership style; and what they *should* do, which is prescriptive. Some theory models are normative, in other words, it is "better" to be at one end of the scale than at the other, or the "normal" is male behavior and attitude with which women are compared. I am not about to give a long description of each theory, but a summary of the more important findings will help you understand current thinking and gain an awareness of how these theories may affect women.

WHAT MOTIVATES PEOPLE

Understanding what motivates people is important, since a manager will act according to what he or she believes will induce the workers to be more productive. Often, the underlying reasons for behaving a particular way are not conscious. Still, most managers could probably come up with a set of beliefs they have about what motivates workers. Two authors have greatly influenced management concepts in this particular area: Abraham Maslow and B. F. Skinner.

Maslow postulated that people's needs could be categorized into several different levels and that the most basic needs had to be realized before the next level could be fulfilled. At the bottom of his *hierarchy of needs* are the physiological needs (food, shelter, sleep, etc.) and safety needs (protection from the threat of danger and deprivation). Only after these have been met can the next order of needs be pursued. These are the social needs: belonging (acceptance, affection, warmth) and self-esteem (status, recognition, respect). At the highest level of the hierarchy is the need for self-actualization, which leads to self-fulfillment, the realization of the full potential and creativity of each individual.

The traditional role for a woman—wife and mother—was in the past seen as fulfilling all her needs: Survival and social needs were met, and she had attained the ultimate in female self-actualization. Today, however, more and more women are striving for self-realization through productive and creative work. Yet, most of these women—many more than men—must put the major part of their energy into meeting survival needs.

Every time a woman walks at night alone in a large city, she is concerned for her *safety* needs much more than men are and is therefore less mobile than men. If women are more frightened than men it is with good reason. Men have been traditionally the protectors of women, and indeed when women are pregnant or nursing or caring for small infants, they are in a more vulnerable condition than men. They cannot run as fast or fight as hard encumbered as they are with children. Women are also more vulnerable to assault and rape, and can be physically overpowered. In other words, women need to worry about their safety more than men do.

Many families are headed by a single woman, who must feed, clothe, nurture, and educate children, care for the home, and maintain full-time employment. If she is lucky, she can fit in her career. Most men do not have to deal with multicommitment, so they can work on fulfilling their needs. And since the "old boy's network" has been around for a long time, they find the belonging and self-esteem needs much easier to fulfill than do women, who must struggle to overcome male discrimination.

The bind many women find themselves in is that self-realization for them encompasses both their personal lives and their professional ones. Giving enough time to each to provide true self-fulfillment can be often problematic. Perhaps what women need to look at is the possibility of arriving at self-fulfillment sequentially. Not everything must be done fully and perfectly at the same time, but some things may take on a high priority during some periods of one's life, and other things come more into focus at other periods. For men, the hierarchy of needs in a linear fashion is much more evident than it is for women.

Skinner's contribution to motivation theory is the idea of behavior modification: People will repeat behavior that is rewarded and avoid behavior that is punished. Of course, rewards and punishments are different for different people; but in its most basic form, the concept has direct implications for women in business. For example, if time after time at meetings a woman's views are ignored or discounted because of sex discrimination, that woman will eventually stop trying to make her opinions known. This not only jeopardizes her own future in management, but the futures of all women: Giving up will strengthen the stereotype (women are passive, they don't understand business, they are weak, they can't deal with pressure or authority, they don't want to get ahead) and therefore the discrimination, as well.

However, if someone comments on what the woman says in the meeting, a subtle reinforcement of her attempt to be heard has taken place. Such comments as "The value of Mary's statement is . . ." or "Building on what Mary just said . . ." or "As Mary pointed out . . ." are verbal acknowledgments of her contribution to the discussion, and they act as encouragement to repeat that behavior in future situations.

While Maslow and Skinner wrote about what motivates people generally, Frederick Herzberg wrote about what motivates workers specifically. He formulated a theory based on personal satisfaction. Motivators, according to him, satisfy and lead to high performance. These are achievement, recognition, the work itself, challenge, variety, autonomy, and promotional opportunity. There are also what he terms *hygiene factors,* which will determine whether you stay on the job but do not push you to perform. These are wages, benefits, the work environment, the type of supervision.

Until very recently, most of Herzberg's motivators were not available to women. Traditionally, women have worked at the very bottom of the business hierarchy, where the jobs are not challenging or interesting, and where there is very little variety, autonomy, or promotional opportunity. So in the past, women have had to concentrate on securing adequate wages and benefits and a pleasing work environment—that is, they focused on hygiene factors, which do not lead to high performance.

David McClelland's research deals with what motivates *successful* people—those who had already made it to the top or were well on their way—and found four common factors:

1. These people were able to set goals and to stretch themselves to achieve those goals.

2. The goals were set within a parameter that the people themselves could influence—that is, goal achievement was not left to chance or within the control of others.

3. The people enjoyed work for the work itself. Success was for its own sake, not for any rewards it brought.

4. They preferred receiving feedback on their performance; they needed to know the score.

He found that these people also spent time thinking up ways of doing things better or more efficiently, a habit started most probably in their childhood.

Of these successful people, McClelland also found that their predominant needs determined the direction of their growth. For instance, those with a predominant need for power were more likely to be executives; those with a need for achievement were more likely entrepreneurs. Those people with a predominant need for affiliation were less successful because they feared disapproval and/or separation from supportive peers. We see here the problems women face, since their needs for approval and affiliation have been traditionally high and their need for achievement and power have not been seen as legitimate.

Douglas McGregor, best known for his Theory "X" and Theory "Y," grouped managers according to their assumptions about workers. Theory "X" managers have a certain set of assumptions and Theory "Y" have an opposite set. Some managers fall into one of the two categories; others fall somewhere in between. The table on p. 273 shows the assumptions of the two different types of managers.

The stereotypical male image of women falls right into place with the Theory "X" manager's beliefs: Women are passive; are not ambitious; are incapable of handling responsibility outside of the home; are more interested in family than in their careers; do not have the same need for achievement and recognition that men have; are not risk takers and therefore are less

Theory "X"	Theory "Y"
Most people prefer to be passive—to be led, to be controlled by others, and to be dependent upon others' decisions.	Most people prefer to be active—to lead, to control their own lives, to be independent, and to make their own decisions.
Most people would rather avoid work.	If work provides satisfaction, most people prefer to do it.
Most people are not very ambitious and either dislike responsibility or are incapable of handling it.	Most people want to get ahead, are capable of assuming responsibility, and find satisfaction in doing so.
Most people cannot satisfy their more important needs through work, and therefore must seek satisfaction outside the work setting.	Most people can find satisfaction in their work and can obtain self-fulfillment through it; work can provide achievement needs and recognition needs.
Most people want security and will resist change.	Most people enjoy the opportunity of challenge and are capable of rising to the occasion.
Most people are not very bright and can easily be misled.	Most people possess creativity as well as a capacity to evaluate, and prefer getting an opportunity to use these talents.

creative and cope poorly with crisis. A manager of female subordinates would then prefer to have strict control over their work, making the assignments, coordinating the activities, distributing materials, setting deadlines, and specifying the behaviors. The women would have narrowly-defined tasks and responsibility for only small segments of the total organization's output. However, the same manager may have a Theory "Y" assumption about the male subordinates.

It is therefore important for women to not only be aware of the way Theory "X" and/or Theory "Y" may be operating in their place of work, but also to be aware of the way their bosses and managers in general may hold Theory "X" assumptions about female employees while holding Theory "Y" assumptions about male workers without being conscious of differentiating.

WHAT MANAGERS DO AND HOW THEY DO IT

Many theorists have focused on managers—on what tasks and responsibilities they must assume, on what roles they must play, on what strategies and tactics they must adopt. Although these theorists have not distinguished the roles as fitting more either first-line, middle, or upper managers, I believe that, depending on the emphasis given to the particular role, the leadership that emerges *will* represent a specific rung of management. Craig Lundberg's theory is a good example of this.

Lundberg sees management in three sequential levels. The first level deals with things, the second level with people, and the third level with ideas. At Level One the concerns are primarily *technical.* One needs to know *how* to do it. These are supervisory positions and many hands-on staff jobs, such as preparing budgets. The supervisors must know the technical aspects of the work their subordinates do so they can train their people. They deal with how the job gets done. At Level Two the concerns are primarily *social.* One needs to know how to deal effectively with people. This is the realm of middle management, those who decide who does what. Ability to delegate is important. At Level Three the concerns are primarily *conceptual.* One needs to be able to deal with abstractions, to analyze, synthesize, to ask questions as well as to answer them. This is the prerogative of the top executive, who must ask "What needs to be done?" Long-range planning comes in at this stage as well as organizational goals.

Where are the women? Women are in the technical aspects of the sequential ladder. They are doing a job, as opposed to pursuing a career. Only 6 percent of working women are at Level Two, although most women would deal quite effectively with the social aspects. Only 1 percent of women are at Level Three, the executive level, where policy decisions are made. Women should therefore seek work in areas that will utilize their interpersonal skills and learn as much as possible about the organization. They should also start thinking in terms of larger company objectives and stop doing only what needs to be done for today without placing it in the context of the whole.

Women who see their work in terms of a job see it in technical terms. Women who see their work as part of a career path are functioning at either the social or conceptual levels. Secretarial work may be seen as a boring dead end job, or it can be an entry to a career.

Henry Mintzberg's studies are based on the close observation for many days of half a dozen male managers. A team of observers wanted to see what, in fact, managers do. From this data, Mintzberg formulated that managers perform three main roles, which he called *interpersonal, informational,* and *decisional,* listing ten major functions for the three categories. Again, although Mintzberg says that a manager performs all of these functions at various times, it seems to me that the focus given each role would be an indicator of management level.

In the *decisional* role, according to Mintzberg, managers are responsible for four main functions: They handle *disturbances* (manage conflict); they *allocate* resources; (who gets what or how much), they *negotiate* for budgets, salaries, and deals; they act as *entrepreneurs,* the innovators, the idea people.

In order to make good decisions, managers need sufficient information. In the *informational role,* managers need to be *monitors* (to gather information), *disseminators* (to share information with the appropriate people), and *spokespersons* (to act on behalf of their subordinates.)

In order to get enough information to make good decisions, managers need sources, which are most often found in the informal network. Here is where managers need interpersonal skills, to gather the necessary data. In the *interpersonal role,* managers are *figureheads* (representing their company to the outside world), *leaders,* (responsible for their workers' productivity and for hiring, training, evaluating, and monitoring,) and finally, *liaisons* (with their peers and with others outside their units or departments).

The importance of all this for women is that most women are not part of an organizational network in which they can use their *interpersonal* skills in order to gain access to *information,* which will permit better *decision making.* The strategy for women would be *at all cost* to start with the informal network, using their interpersonal skills in order to become first monitors, then disseminators, then spokespersons, to be able to move on through *all* of Mintzberg's roles and functions.

WHAT MANAGERS SHOULD DO

It is important to distinguish between contingency theories and normative theories of leadership. Normative theories suggest that there is a "best way" to manage. Contingency theories say that there is no single ideal management style that will fit all people in all situations all of the time—a situational leadership style is recommended. Rensis Likert, Robert Blake and Jane Mouton, and Chris Argyris have all developed normative leadership theories; Fred Fiedler and Paul Hersey and Ken Blanchard suggest a contingency style.

Rensis Likert established a hierarchy of management systems that were descriptive of basic styles of management, and he also ranked them according to effectiveness. A System I manager is *exploitative-authoritative*, which is self-explanatory. A System II manager is *benevolent-authoritative*, a sort of paternalistic taskmaster. A System III manager is *consultive*, which means that he or she checks with others but makes the final decision alone. Finally, a System IV manager is participative—decision making is shared with subordinates whenever feasible. According to Likert, this is the ideal system; it establishes an atmosphere of trust among employees that leads to high performance and more effective leadership.

Likert's second theory, the linking pin theory, discusses the fact that people in an organization belong to more than one group in an ascending order. As you see in the drawing below, each person at the top of a small pyramid also belongs to the next higher pyramid. In other words, a first-line supervisor belongs to both the workers' group and to a group made of other first-line supervisors and a middle-manager, who in turn belongs also to a group of upper-level executives.

The Linking Pin Theory

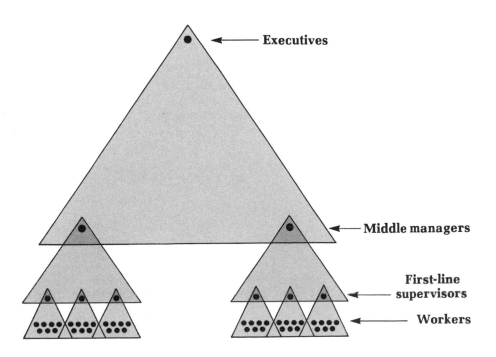

Executives

Middle managers

First-line
supervisors

Workers

It is important for women to be aware that, as members of the organization, they belong to several groups, and that they need to participate as active members in each. Not only must women be competent as managers, but they must also be competent as colleagues of other managers. It is therefore critical for women to be included in any peer group activity. An organizational chart that pinpoints your spot, using Likert's linking-pin diagram, may help you understand your position in terms of responsibilities and opportunities below, above, and across.

As you can see from the diagram, supervisors or managers are successful only if they can influence the level above them and be part of that group. In other words, to be successful they not only influence their subordinates, they also influence their superiors. They represent their units or departments to upper management. That is where they go to bat for their people. Successful managers must be seen as credible, as able to defend the budget, as having access to resources, and as able to get promotions and salary increases for their employees.

Robert Blake and Jane Mouton evolved a grid (see below) that asked people to rate themselves on a continuum from all-concern for production to all-concern for people.

The Managerial Grid

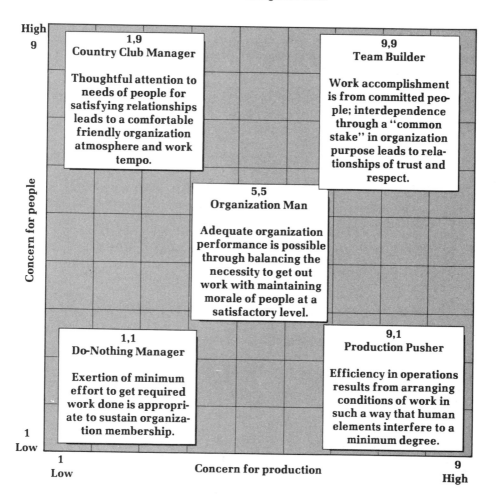

Every manager, of course, has at least one back-up leadership style, sometimes two or three, depending on the situation and/or the people involved. Some managers hopscotch around the grid, jumping from one grid position to another. Managers jumping all over the grid, depending upon the particular situation confronting them, convey to their subordinates the absence of a consistent way of behaving which, I may add, is very disturbing to subordinates who feel that the boss is handing out differential treatment.*

We believe that the 9,9 style of leadership *is* the ideal approach to all leadership situations. This means that a strategy of relationship premised upon mutual respect, confrontation of differences, mutually formulating goals and objectives, learning from mistakes through critique, etc. is certainly ideal; and the last fifty years of research on leadership has verified that when this mode of leadership is exercised, the consequences include higher productivity and greater satisfaction. It is also now becoming clear that consistent behavior of this sort is also associated with physical and mental health. We have not said that a manager is necessarily able to apply the 9,9 leadership style in all situations. That is contradictory to common sense.

Part of the confusion here may be related to distinguishing between strategy and tactics. The distinction is between strategy, which is related to the underlying fundamental premises that orient conduct, and tactics, i.e., the specific behaviors that are selected consistant with that strategy to deal with particular situations. Thus, a 9,9 orientation is vastly different whether one is dealing with a new employee or an old tried-and-true hand. A 9,9 orientation may remain constant, but the tactics of its application vary with the needs of the situation. What is maintained is trust and respect, openness, candor, confrontation of differences, etc. What is shifted is the amount of time spent with the subordinate, the extent to which it is necessary or desirable to use critique for the purpose of aiding learning . . . , setting goals, or, alternatively, being aware that . . . the old hand already is well geared to achieving relevant corporate goals. All of these specifics vary under a constant orientation.†

What is important is the skill to diagnose the situation and know the people's needs and tolerance levels. This takes good judgment and a repertoire of behaviors that allows the manager to respond correctly. Since the more successful managers are those who can combine both production and people concerns, women should be careful to balance both. To overcome the stereotypes about women, some successful women have at times bent over backwards to not seem overly people-concerned and have then become overly production-concerned.

This differentiation between strategy and tactics is important for women in executive positions to understand. The strategy remains constant, but within it the tactics may change.

Chris Argyris put his finger on something that happens everywhere, all the time, but is seldom confronted adequately. Simply put, his "espoused theory" is what people say they believe; his "theory in practice" is what people actually do. It is therefore important for women to see what organizations *say* about hiring and promoting women and then look at what is actually *done* about it. At what levels are the women in the organization? Are they indeed being given opportunities for advancement? Are women in the training programs? Some employers say everyone will be given an equal opportunity, but women are hired for only

---

*Robert Blake and Jane Srygley Mouton, "Should you teach there's only one best way to manage?" *Training, the Magazine of Human Resources Development* (Minneapolis, MN: Lakewood Publications), April 1978.

†Dr. Robert Blake, in a personal letter, March 23, 1979. Reprinted by permission.

entry-level positions and then remain there. Therefore, the espoused theory would be that everyone is treated equally; the theory in practice then belies it.

According to Fred Fiedler, the decision as to which leadership style is most effective rests upon three characteristics. As these characteristics change, so should the leadership style. Fiedler's three characteristics are: (1) the power of the leader; (2) the quality of relationship between the leader and those being led; (3) and how much certainty or uncertainty there is in the task. For example, in an assembly line where there is task certainty, authoritarian leadership may work well and relationships do not have to be good between supervisor and workers. However in a research and development unit where there is task uncertainty, leadership has to be collegial and based on good relationships.

Women have traditionally had difficulty obtaining inclusion in male collegial groups. The quality of the total work may suffer from the lack of performance feedback women receive from their male peers and bosses. It is very important in potentially collegial situations to be sure that as a woman, you are included. In situations where your relationship with your boss is critical to the job, it might be best to leave if your relationship seems hopeless. Women who are new to the world of work may feel at a disadvantage in a case of task uncertainty. In all beginnings, uncertainty is threatening because there is as yet so little security. It is only with a long time at a particular place of work that both men and women begin to feel that they can cope with uncertainty, deal with challenge, and welcome variety. Therefore it is important for women to know their capacities for task certainty/uncertainty before accepting certain jobs that have a high element of unpredictability.

Other theorists have used different variables to determine a contingency management style. Paul Hersey and Ken Blanchard developed a model based on subordinate maturity (task ability and willingness to learn), which states that the more mature the subordinate, the less authoritative the manager has to be. Victor Vroom and Philip Yetton formulated a theory based on the characteristics of the problem at hand. Depending on the importance of the decision, the information available, the commitment of the subordinates to organizational objectives, and several other variables, Vroom and Yetton proposed five varying leadership styles, from autocratic to shared decision making, for handling any situation.

OTHER STUDIES

In addition to these theories there are two studies that also stand out as part of the knowledge base, known as the Hawthorne Effect and the Rosenthal Effect.

The Hawthorne Effect refers to an experiment undertaken by Elton Mayo at the Hawthorne Western Electric Plant in the 1920s. Groups of workers were observed under a variety of conditions, and the results were surprising. No matter how the working environment changed—that is, for better or worse—the worker's productivity increased. The study indicated that the simple fact of being observed by outsiders, of "being in the limelight," spurred people to high performance. Token women in the limelight should be aware of the impact this public scrutiny may have on them. The added visibility will place performance pressures on some women and make others "camera-shy."

The Rosenthal Effect refers to studies done in the late 1960s by Robert Rosenthal. Although Rosenthal's experiment design was later criticized, many theorists still consider the results to be of value. School teachers were given their students' IQ scores at the beginning of the year. At the end of the school year, when all the children were tested again, it turned out that those children believed by the teacher to be brilliant did in fact excel, while the children considered to be below average turned in below-average test results. What was not told to the teachers was that the original IQ scores were an approximation of the children's locker numbers, not a reflection of their intelligence. The brilliant and less-than-average scores had been selected randomly. The teachers' *expectations* of superior and inferior performance produced just that.

This demonstrates that not only will high expectations produce better results but that the opposite is also true. In the face of bosses who assume that women cannot do the job as well as men, we women need to believe in our own abilities so that *our* expectations—*our* predictions—come true, not theirs. We can make a real *self*-fulfilling prophecy, not another person's-fulfilling prophecy, come true for us.

There are two routes to success in business: formal, through schooling, and informal, through experience. Many people who have climbed the corporate ladder have not had the advantage of a formal education, such as a business degree or an MBA, but it is my experience that these people often have a gnawing feeling that they lack some legitimacy, that they got in through the back door. Many women have admitted they are afraid of being "found out"—found out that they don't know something basic that they *ought* to know. Just what that something is, is not always clear, but I suspect that it has to do both with early exposure to business and with a theoretical framework, which is generally discussed only in a classroom situation. Although business colleagues will probably refer to the concepts only on rare occasions, the descriptions of the predominant theories of management offered here should provide you with a frame of reference and therefore the self-confidence that stems from both knowing and knowledge to feel more comfortable when these are mentioned. If you would like more complete information, the following readings are suggested:

Argyris, C., and Schon, D. A. *Organizational Learning: A Theory of Action Perspective.* Reading, MA: Addison-Wesley, 1978.

Blake, Robert R., and Mouton, Jane S. "Interview." *Group and Organization Studies* 3–4(1978).

____. *The Managerial Grid.* Houston, TX: Gulf Publishing Co., 1964.

Hampton, David R. *Behavioral Concepts in Management.* Los Angeles: Dickenson Publishing Co., 1972.

Hersey, Paul, and Blanchard, Kenneth. *Management of Organizational Behavior* (third ed.). Englewood Cliffs, NJ: Prentice-Hall, 1977.

Herzberg, Frederick. *Work and the Nature of Man.* New York: World Publishing Co., 1966.

Herzberg, F.: Mauser, B.: Snyderman, B. *The Motivation to Work.* New York: John Wiley & Sons, 1959.

Homans, George C. *The Human Group.* New York: Harcourt, Brace & World, 1950.

Kolb, David A.: Rubin, Irving M.; McIntrye, James M. *Organizational Psychology: An Experiential Approach* (second ed.). Englewood Cliffs, NJ: Prentice-Hall, 1974.

Likert, Rensis. *The Human Organization: Its Management and Value.* New York: McGraw-Hill, 1967.

Maslow, Abraham. *Motivation and Personality* (second ed.). New York: McGraw-Hill, 1970.

Mayo, Elton. *The Human Problems of an Industrial Civilization.* New York: Macmillan, 1933.

McClelland, David. *The Achieving Society.* Princeton, NJ: Van Nostrand, 1961.

McGregor, Douglas. *The Human Side of Enterprise.* New York: McGraw-Hill, 1960.

Miles, R. E. *Theories of Management: Implications for Organizational Behavior and Development.* New York: McGraw-Hill, 1975.

Mintzberg, Henry. *The Nature of Managerial Work*. New York: Harper & Row, 1973.

Rosenthal, Robert. "The Pygmalion Effect." *Psychology Today*, January 1973.

Skinner, B. F. *Contingencies of Reinforcement*. New York: Appleton-Century-Crofts, 1969.

Stewart, Rosemary. *Managers and Their Jobs: A Study of Similarities and Differences in the Way Managers Spend Their Time*. London: Macmillan, 1967.

Vroom, Victor, and Yetton, Philip. *Leadership and Decision Making*. Pittsburgh: University of Pittsburgh Press, 1973.

Wren, Daniel A. *The Evolution of Management Thought* (second ed.). New York: John Wiley & Sons, 1979.

Appendix F: Social Security Pensions

As specified by the Social Security Administration Office of Research and Statistics, "a woman becomes fully insured if she has at least one quarter of coverage for each calendar year elapsing after 1950, or if later, after the year in which she attained age 21, and prior to the year in which she attained age 62."

I will show you how I calculated this for myself. In 1950 I was 24 years old. From 1950 to 1989 when I'll be 62, thirty-eight years will have elapsed. In order to qualify for social security I need to have worked one-quarter of these thirty-eight years, or nine and one-half years. I can then either collect 100 percent of my own benefits or fifty percent of my husband's, whichever is larger (laws change rapidly; check with your local social security office for the latest update).

In 1900, 4 percent of the U.S. population was over 65 years of age. By 1960, it more than doubled, to 9 percent, or 16.6 million people. The Department of Health, Education, and Welfare projects that this number will almost double again to about 30.6 million by the year 2000. Therefore, there will be countless opportunities to work in areas related to the aged. Most universities have gerontology studies; these might be places for you to gather more information.

General Bibliography

Blaxall, Martha, and Reagan, Barbara (Eds.). *Women and the Workplace*. Chicago: University of Chicago Press, 1976.

Boserup, Ester. *Woman's Role in Economic Development*. New York: St. Martin's Press, 1970.

Cherry, Rona B., and Cherry, Lawrence B. *The World of American Business—An Introduction*. New York: Harper & Row, 1977.

Gordon, Francine, and Strober, Myra. *Bringing Women into Management*. New York: McGraw-Hill, 1975.

Gornick, Vivian, and Moran, Barbara K. *Women in Sexist Society—Studies in Power and Powerlessness*. New York: Mentor Books, 1971.

Hall, Francine S., Gudwer, Rosanne D., Sapp, Kathryn F. *Myth, Ms., Mgr.. A Bibliography For and About Women in Business*. Kenosha, WI: University of Wisconsin Press, 1978.

Harragan, Betty Lehan. *Games Mother Never Taught You (Corporate Gamesmanship for Women)*. New York: Rawson Associates, 1977.

Hennig, Margaret, and Jardim, Ann. *The Managerial Woman*. New York: Anchor Press/Doubleday, 1977.

Josefowitz, Natasha. "The Clonal Effect in Organizations." *Management Review*, September 1978.

Kanter, Rosabeth Moss. *Men and Women of the Corporation*. New York: Basic Books, 1976.

Kanter, Rosabeth Moss, and Stein, Barry A. (Eds.). *Life in Organizations*. New York: Basic Books, 1979.

Larwood, Laurie, and Wood, Marion M. *Women in Management*. Lexington, MA: Lexington Books, 1977.

Loring, Rosalind K., and Otto, Herbert A. *New Life Options—The Working Woman's Resource Book*. New York: McGraw-Hill, 1976.

Loring, Rosalind K., and Wells, Theodora. *Breakthrough: Women into Management*. New York: Van Nostrand Reinhold, 1972.

Miller, Jean Baker. *Toward a New Psychology of Women*. Boston: Beacon Press, 1978.

Rohrbaugh, Joanna Bunker. *Women: Psychology's Puzzle*. New York: Basic Books, 1979.

Stead, Betty Ann. *Women in Management*. Englewood Cliffs, NJ: Prentice-Hall, 1978.

Index

Accessibility, 16–17, 18
Accountability, 16, 18, 75, 159, 174, 213
Adolescents, 134
Advancement. *See* Upward mobility
Affiliation needs, 17–18
Affirmative Action, 45, 55, 90, 211
Age: for childbearing, 126, 132; of children, 132, 133–134; discrimination, 40, 229, 232; and life cycles, 125–126, 230; older women, 34, 36, 47, 126, 231–233; retirement, 229; of sponsors, 97; stereotyping, 53, 126, 229, 232
Aggressive behavior, 84, 118
Alliance Against Sexual Coercion, 55
Analytical ability, 26
Anger, 55, 81, 118–119; style of expression, 55
Apprentice, defined, 93
Aspirations, 65, 77, 128, 212
Assertiveness, 15–16, 18, 83–85; strategies for, 85–88
Assessment Center, 38
Attitude, negative, 121–122
Attributed power, 6–7, 159–160
Authority, 15, 18, 105, 166; defined, 199; and male subordinates, 108, 173–174

Beckhard, Richard, 222
Bennis, W. A., 181
Boredom, 88–89, 155
Bosses, 83; evaluation by, 167–168; evaluation of, 120; knowing expectations of, 66, 83, 110, 166–168, 169; as mentors, 95; sexual harassment by, 54; women as, 108
Boston area, 195
Brody, Gerald D., 37n
Budget, 75–77, 79, 145–146

Burnham, D. A., 200
Business, starting your own, 231–232
Business and Professional Women's Club, 101

Career: defined, 128n; family and, 5, 16, 51, 75–76, 98, 125–126, 127, 131–133; goals, 65, 128, 144; his and hers, 130; interrupted or delayed, 131–133; versus job, 128
Career mobility, 17, 24, 130, 131
Child care, 133, 135–136
Children, 8; age of, 132, 133–134; day care for, 133, 135–136; decision to have, 126, 131–133; in two-income families, 128, 131–133
Christie, Agatha, 232–233
Clothing, 35, 39, 45–46, 51, 56, 63, 73
Colleagueship, 168–170
Cohen, Allan, 191n
Committees. *See* Groups; Meetings
Communication, 171–173
Competencies, 25–27, 28–29, 68; mastering skills, 77–81; for top management, 199
Competition, 168, 189, 223
Conflicts: career, 130, 131; in groups, 223–225; in multicommitted families, 125, 127, 129–130, 133. *See also* Stress
Conformity, 56–59. *See also* Norms
Content, 10, 11
Coordinating ability, 26
Coping skills, 133–134, 137–138, 155–157, 200
Creativity, 87, 201
Credentials, 25–27
Credit, giving, 109
Credit discrimination, 135

Crisis management, 76
Criticism: constructive, 17, 66–67, 109, 121, 172; personalizing, 17–18, 30, 149, 167; requesting, 167–168; withholding, 108
Cuba, sharing housework in, 133n

Day care, 133, 135–136
Decision-making, 76–77, 87, 119; and access to information, 75; in management, 146–149, 160; in marriage, 130; sexual differences in, 10, 146, 148, 153; steps in, 147
Defensiveness, 166
Delegation of responsibility, 16, 76, 109, 117–119; in groups, 183; in middle management, 152, 171–175, 201
Dependency: versus entitlement, 11–12; in groups, 180; homemakers, 126, 127
Dependent power, 8, 82, 91
Depression, 81–82, 118–119, 154
Discrimination. See Age, discrimination; Sex discrimination
Disengagement, in employees, 118–119
Doctor's office, resources in, 112
Double bind, 15, 15n
Double bosom theory, 14
Dual-career families, 125–126; career competition, 130, 131; mobility, 130–131; parenting, 131–134; practical and emotional responsibilities, 128–129; sharing two incomes, 126–134

Eddy, Janice, 99n
Effectiveness, power, 4
Elderly, 229, 231, 232
Emergencies, 152, 200
Emotionality, 55–56, 167–168
Emotional responsibility, in marriage, 129
Employment agencies, 90
Empowerment, 5, 12, 18, 199
Entitlement, 11–12, 18–19, 130, 160, 207
Entrepreneurship, 201, 231
Entropy, in systems theory, 221
Entry-level position, 51–68, 82
Equal Credit Opportunity Act of 1975, 135
Equal Pay Act of 1963, 135
Europe, day care in, 133
Evaluation, performance, 119–121, 172

Executive level: accountability, 213; benefits and costs of, 11, 214; collaborations, 223, 224; functions of, 213–219; issues, 218, 219–225; mobility trends, 130; spheres of influence, 218–221; stereotypes, 72–73; trust relationships, 213, 217–218; turnover rates, 88. See also Leadership; Upward mobility
Exempt and nonexempt, 44, 44n, 77
Expectations: of a boss, 66, 83, 110, 166–168, 169; dealing with roles, tasks, and resources, 73–77; income, 25, 65; job, 47–48, 59–61; sex differences in, 5, 15–16, 67, 74
Expression, fear of, 86–87, 88
Extrinsic reward, 118
Eye contact, 80

Family. See Dual-career families; Multicommitment
Feedback, 167–169, 172–173, 182, 185
Financial management: at executive levels, 220; in marriage, 27, 130, 137
Financial manager, 31
Fink, Stephen, 191n
Firing, 6, 121, 122–123
First impressions, 52
Fishman, Pamela, 81
Followers, 199–200
Forcefulness, power, 4
Forum for Executive Women, 101
Friendships, 17–18, 25, 35, 170; friendly versus familiar, 52, 160–162, 168

Gadon, Herman, 191n
Gantt Chart, 113
Gesture initiation, 87, 184
Goals: career, 65, 128, 144; income, 25, 65; making them known, 95–96; personal, 144, 149–152, 155, 208, 231
Gossip, 167; as a resource, 82, 101
Government agencies, 215–217
Group interview, 38
Groups: competition between, 223–225; consensus, 225; control and direction, in, 187–192; interdependence in, 179–180, 223; intergroup conflict, 223–225; leadership in, 30, 74, 177, 182–191; participation in, 11, 30, 177, 179–185, 190–191; phases of development, 179–

187; seating arrangements, 186–187; sexist behavior in, 181–183; sexual differences in, 177–178, 181, 188

Hall, Douglas, 128
Haskell, John, 153n
Health, 138, 156, 201, 230
Health services, 217
Hinckley, Stan, 161n
Hiring practices, 5, 42, 44, 116; clonal effect, 45–47; as a supervisor, 113–116
Hobbies, 35
Homemakers: dependency of, 126, 137, 232; marketable skills, 27, 34; stereotypes, 40
Homosexuality, 136
Horizontal moves, 24, 53
Housework, sharing, 128–129, 133n
How to Get Control of Your Time and Your Life (Lakein), 149–150
Humor: organization norms, 56; in public speaking, 80; with sexual overtures, 54, 163; in stressful interviews, 43

Identity, in supervision, 103, 107, 170
Inclusion, need for, 52
Income: dual-career families, 126–134; executive level, 195; goals and expectations, 25, 65; learning office policy on, 66; retirement, 229; sexual discrimination, 65; single women, 135
Independence, 126, 134–137, 179–181
Influence power, 8–10, 76; through reciprocity, 82
Information: access to, 75–76, 82–83, 101, 147, 148, 163, 206; current, 78; transmitting, 78–79, 166
Inhibition, 83, 84, 85
Initiative, 96
Instructional ability, 26, 109
Interdependency: in groups, 179–180; 223; power of, 8, 168, 169
Intergroup skills, 30, 223
Intermediate power position, 8, 91
Interpersonal skills, 30
Interviewing, guide to, 114–115
Interviews: for evaluation, 120–121; preparing for, 32, 36–39; questions to ask, 43–45, 46–47; stress, 41–43, 44; types of, 38, 39; typical questions in, 39–41

Intrinsic reward, 117
Introductions, problems with, 136n
Intuition, women's, 10, 67, 108, 152, 220
Israel, women in, 128n

Job application: credentials and competencies, 25–31; interviewing, 35–45; letters of inquiry, 32, 35; preparation, 23–24, 29, 37; résumés, 33–34, 90
Job titles, 8, 44, 66, 73, 81
Jobs: application for (see Job application); choosing, 24–25; dead-end, 88–89; expectations, 47–48, 59–61; first, 51–59, 61; fitting workers to, 110–113, 122; norms, 56–59; references, 90, 123; satisfaction, 24, 65, 89, 118–119, 123; searching for, 31–45; second, 90–91
Josefowitz, Nina, 83n
Jung, Carl, 231

Kanter, Rosabeth Moss, 14, 195–196
Knowing and knowledge, 10–11, 67, 220–221
Kubler-Ross, Elizabeth, 19n

Lakein, Alan, 149–150
Language use, 78–81; de-sexing, 80, 81
Latch-key children, 132, 132n
Laws, affecting women: in job application, 34, 42, 43, 113; reference sources for, 43; with sexual harassment, 55
Leaders, 9, 57
Leadership: defined, 199; issues, 206–208; qualities of effective, 16, 196, 200–204; sex differences, 202–203; style, 204–206. See also Executive level; Groups, leadership in
Leisure time, scheduling, 129–130, 133
Letters: after interviews, 36, 42, 43, 44; of inquiry, 32, 35
Levin, Barbara, 36n
Ley, David, 117n
Life cycles, 125–126, 230
Life-styles, alternative, 136
Living together, 136
Loneliness, 13, 105
Loyalties, divided, 105
Luck, 220

McClelland, D. C., 200
McKinnon, Jennifer, 24
Management by objectives (MBO), 119
Management functions, 26, 27, 31
Management Recruiters International, 115n
Managers: accessibility, 16–17; affiliation dependence, 7; emotionality, 55; formality and informality with, 57; risk taking, 67; potential, 70, 72–73; skills needed, 79, 217; women versus men, 74. *See also* Executive level; Middle management
Manpower, 90
Manufacturing operations, 216
Marketable skills, 26, 27, 28–29, 34
Marketing, 31
Marriage: competing careers, 130; division of labor, 128–219, 133; dual careers, 131, 133, 134; interdependence, 8, 128–129, 134; one career, 137; practical and emotional responsibilities, 129
Married men, new trends with, 130
Married women, sexual harassment of, 54
Math skills, 77, 79
Meetings, 11, 151, 166, 167. *See also* Groups
Mentor, 93–95; defined, 93
Middle management: accountability, 159, 174, 213; colleagueship, 168–170; decision-making, 146–149, 160; delegating, 152, 171, 173–175, 201; functions of, 142–143, 219; giving and requesting feedback, 167–168, 169, 172–173; information networks, 147, 148, 160, 163, 166–167; managerial style, 148, 163–166, 168, 183; personal skills needed, 144–157; power balances, 159–162, 166, 169; professionalism, 162–163, 168; risk taking, 145–146; social relationships, 152, 162–163, 168–170; stress in, 154–157; superior-subordinate relationships, 159–162, 163, 166–175; time management, 149–154
Miller, Jean Baker, 195
Mintzberg, Henry, 154
Mobility: in dual-career families, 130, 131; sexual differences, 34
Monteiro, Lois A., 65
Motivation, 117–119
Mulins, Maruta, 112
Multicommitment: conflicts, 129–130,

131, 154; continuing education, 125, 127; coping skills, 133–134, 137–138; defined, 125; family and career, 5, 16, 51, 96, 98, 125–133; leisure time, 129–130, 133; new trends in, 130–131, 196, 208; personal control areas, 137–138; priorities, 215, 218; single mothers, 51, 135, 136. *See also* Dual-career families
Murningham, Marcy, 15n

New Girls' Network, 101
"New women's networks," 101
Nonconformity, 56, 59
Norms, job, 56–59, 61, 71; nonconformity as, 59; well-defined and ill-defined, 57; within departments, 106
Notice, giving, 90

Objectives. *See* Goals
Objectivity, 122
Occupational fathering, 95, 96
Office, location of, 62, 76, 86
Organizational skills, 30
Organizational structure, 30, 31; formal and informal, 61–64; learning about, 64; systems theory, 221–222
Organization behavior (OB), 222
Organization development (OD), 222
Organizations: investigating, 38–39; kinds of, 25
Orientals, 18
"Out of synch," 126, 126n

Panel interview, 38
Parents without Partners, 136
Part-time work, 127, 136, 137
Passive behavior, 84, 118
Paternity leave, in Sweden, 133n
Performance evaluation, 119–121
Perqs, 66
Personal goals, 144, 149–150, 152, 155, 208, 231
Personal matters: on resumes, 34; at work, 52, 160–161
Personal skills, 30, 144–157
Personnel department, 35, 36, 89
Politics, 64, 83, 163
Power: attributed, 6–7, 159–160; through

awareness, 3, 18–20; defined, 3–4, 199; dependence versus entitlement, 11–12, 19; dependent, 8, 9, 11, 91; hierarchy, 8–11; influence, 8–11, 76; intermediate, 8, 9, 11, 91; knowing and knowledge, 10–11; in leadership, 203–204; versus powerless, 7; roadblocks to, 5, 15–18; sexual differences, 5, 6, 10, 12–15, 160–161; sharing, 4, 166; sponsors, 95; symbols of, 62–63
Practical responsibility, in marriage, 129
Precedent setting, 66–68
Priorities. *See* Multicommittment
Problem individuals, 122–123, 188
Process skills, 10, 11, 171, 185, 214
Production, 31
Professionalism, 97, 162–163, 168
Professionals, 8, 98
Project planning, 113
Promotion: departmental, 106; desire for, 6; discrimination in, 5, 73; finding sponsors and mentors, 94, 95–96, 166; influential people, 82–83; new identity, 105; as reward, 118. *See also* Upward mobility
Protégée, 93, 94–95, 96, 102
Protestant work ethic, 98
Punitive approach, 121–122

Quality of life, 130, 156–157, 198
Queen Bees, 98–99
Quitting, 88, 89–90

Raises, 118
Reading skills, 77, 78, 153
Reciprocity, 82, 90, 94–95
References, job, 90, 123
Refusing, at work, 87–88
Reports, 151–152
Requesting, at work, 87, 88
Resources, use of, 73, 75–77, 81–83
Respect, 108, 109, 121
Responsibilities. *See* Delegation of responsibilities; Multicommitment
Résumés, 33–34, 90
Retirement, 229–230; options, 230–231
Reward system, 119–121
Risk taking: activities, 87; in groups, 183; in middle management, 67, 145–146,

149; as sponsor, 96; in top management, 201, 220; by women, 59
Roberts Rules of Order, 80
Roles: in a group, 177; learning versus dependent, 52, 53. *See also* Multicommitment; Sex roles

Samuelson, Paul, 52n
Scheduling, 111–112
Schein, Ed, 199
Schutz, William C., 181, 184n, 186n
Second job, 90–91
Secretaries, 63, 74, 83, 163; recording, 184
Self-awareness, 3, 18–20
Self-confidence, 32, 67–68, 96, 145, 183, 189, 196–197, 200
Self-esteem, 65, 67–68, 84, 105, 145
Seniority, 110
Service organizations, 216
Sex discrimination: in credit, 135; in earnings, 65; at executive levels, 211–212; in hiring, 5, 45, 46–47, 155; patterns of, 13, 97, 155, 159; in promotions, 5, 73; "token" women, 155, 180, 211
Sexism: in interviews, 42–43; in language, 81; by women, 98–99
Sex roles, traditional, 10, 13–17, 55, 74, 96–97, 128, 160, 169. *See also* Socialization; Stereotyping, sexual
Sexual attraction, 97, 162–163
Sexual differences: in decision-making, 10, 146, 148, 153; in emotionality, 55–56; in group behavior, 177–178, 188–189; in leadership, 202–203; in paths to power, 5, 6, 10, 12–15, 160–161; in retirement, 230, 231; in self-esteem, 67, 77; in social behavior, 17, 81, 162–163; in socialization, 12–15
Sexual overtures, 53–55, 135, 163, 168
Sexual stereotyping. *See* Stereotyping, sexual
Sharing: in marriage, 128–129, 133n, 134; power, 4, 10, 166
Single women: alternative life-styles, 136; discrimination against, 135–136; mothers as, 51, 135–136, 137; rewards for, 134–136
Skills: coping, 133–134, 137–138, 155–157, 200; marketable, 26–29; personal 30, 144–157; supervisory, 106–107; technical, 77, 141

Smoking, 56
Social consciousness, 218, 219
Socialization: changing the system of, 5, 14–20; female, 12–15, 85, 195–196; male, 12–15, 61, 159, 196
Social obligations, 128
Socioeconomic class, 46–47
Speaking skills, 77, 79–81, 86
Sponsor, 93–99; becoming a, 102, 225–226, 231; defined, 93; finding, 96, 97, 102; functions, 95; males as, 95–97; Queen Bees, 98–99; women as, 97–99
Sponsor-protégée relationship, 93, 94–95, 102
Stereotyping, 2; age, 53, 126; clonal effect in, 45–47; dealing with, in interviews, 39, 40; of executives, 72–73, 202; of homemakers, 10; of recent graduates, 40; sexual, 53, 54, 59–60, 96, 97, 135; through socialization 2, 12; students, 40
Stress: coping with, 154–157; in interviews, 41–42; with retirement, 230. See also Conflicts
Students: expectations, 47; marketable skills, 26; married, 125, 127; preparing for first job, 51; stereotypes, 40
Style: in leadership, 204–206; in management capacity, 148, 163–166, 168, 183; sexual differences in, 6, 163–166; in speaking, 80; in supervisory capacity, 107–110, 119, 122
Subordinates, 170–175, 199; with alternative life-styles, 136; communication with, 160, 206; female, 225–226; leadership with, 203–204, 206; males, 108, 120, 169, 173–174; treating as "staff," 109, 121
Subordination, contingency style of, 83
Success, learning from, 118–119
Successive interviews, 38
Supervision, 6, 105–123, 142; delegation and motivation, 117–119; female, 116; hiring, 113–116; matching workers and jobs, 110–113, 122; setting a style, 107–110; skills for, 106–107, 141
Support systems: for decisions, 121, 160; groups, 101; "new women's" networks, 101; for single women, 136, 137; for stress, 155–157; types of, 93, 99–102
Sweden, paternity leave in, 133n
Systems theory, 221–222

Tannenbaum and Schmidt Leadership Model, 205–206
Telephone calls, 36, 151
Territory, 62, 75, 76, 86
Therapeutic approach, with problem employees, 121, 122
Time: as a resource, 75, 76–77, 133; saving, 153–154, 155; scheduling, 133, 149–154, 156; wasting, 150–153
Time autonomy, 63
Training, 116
Transference, theory, 85, 202
Transfers, 89
Tuckman, B. W., 181
Turnover rates, 88

Unemployment, 90
Unions, 82, 122
Upward mobility: barriers to, 212; becoming assertive, 83–88; choosing, 64–66, 71–73, 196, 198, 212; controlling forces, 73–77; dead-end jobs, 88–89; mastering the skills for, 77–81, 199; ploys of the powerless, 81–83; rules in, 198–199; stress of, 155, 157; transferring, 89

Veiga, J. F., 207n
Venture capital, 231, 231n
Vertical moves, 53
Visitors, social and professional, 152
Volunteer activities, 126, 137

Warden, Dean Charles, 206n
Way, F. W., 195
West, Candace, 81
White male culture, 212, 212n
Willits, Robin, 191n
Withdrawal, as a strategy, 88
Women's movement, 13
Women in Business, Inc., 101
Women's Forum, 101
Women's Lunch Group, 101
Work atmosphere, 25
Work hours, flexible, 136
Working Women United Institute, 55
Writing skills, 77, 78–79, 152

Zimmerman, Donald, 81